WEBSITE GUIDE

For Agribusiness and Horticulture Concepts

Roby Jose Ciju

Author's Note

This book is a compilation of numerous databases on various agribusiness and horticulture concepts. All the databases illustrated in this book were originally sourced from URL: http://www.alexa.com, a leading website providing access to hundreds of databases both category wise and country wise on different topics.

I have ensured that I have accessed databases of almost all agribusiness and horticulture concepts available at alexa.com for writing this book. In fact, going through such an enormous amount of data, and compile them was indeed a nerve-wrecking experience, but I persisted on my mission to deliver a top quality product.

My vision is to make accurate and quality information on various agribusiness and horticulture concepts accessible to all interested parties at their finger tips at an affordable cost. Hope I will be able to do more to fulfil my vision in times to come. If at all any errors are crept in and noticed by my readers, I kindly request to point them out to me at the earliest so that I can rectify them in upcoming versions of the book.

Roby Jose Ciju

CONTENTS

COUNTRYWISE TOP WEBSITES

Top 10 Global Websites

Website URL	Description
Google.com	World's largest internet search engine that allows its users to search for any of the world's information. Provides direct access to Google Videos and Google Images
Facebook.com	World's largest social networking site that connects people
Youtube.com	YouTube is world's largest collection of high quality videos
Yahoo.com	World's second largest internet search engine
Baidu.com	World's third largest internet search engine and also the leading Chinese language search engine
Amazon.com	Amazon.com is world's largest online retail company
Wikipedia.org	World's largest free encyclopedia available online which is built collaboratively by using wiki software
Taobao.com	Taobao is world's second largest online shopping destination
Twitter.com	World's second largest social networking and microblogging site
Qq.com	China's largest and most used Internet service portal owned by Tencent, Inc

Source: Alexa.com

Top 20 Websites in India

Website URL	Description
Google.co.in	Indian version of popular search engine Google.com
Google.com	World's largest internet search engine that allows its users to search for any of the world's information. Provides direct access to Google Videos and Google Images
Facebook.com	World's largest social networking site that connects people
Youtube.com	YouTube is world's largest collection of high quality videos
Yahoo.com	World's second largest internet search engine
Flipkart.com	Flipkart is India's largest online shopping destination
Wikipedia.org	World's largest free encyclopedia available online which is built collaboratively by using wiki software
Blogspot.in	A popular blogging site in India
Indiatimes.com	It is a major information source for a variety of topics
Linkedin.com	India's largest professional networking site
Amazon.in	Indian version of Amazon.com, world's largest online retail company
Snapdeal.com	Snapdeal.com is one of the largest online marketplace in India
Twitter.com	World's second largest social networking and microblogging site
Amazon.com	Amazon.com is world's largest online retail company
Jabong.com	India's leading online shopping destination
Wordpress.com	Free blogs managed by the developers of the WordPress software
Rediff.com	India's leading online portal with free e-mail and many other services
Stackoverflow.com	A language-independent collaboratively edited question and answer site for

	programmers
Pinterest.com	Pinterest is an online place where users can post collections of things they love
Hdfcbank.com	Official website of one of India's leading commercial banks. Information about services offered to resident Indians, NRI's, and companies

Source: Alexa.com

Top 20 Websites in the United States

Website URL	Description
Google.com	World's largest internet search engine that allows its users to search for any of the world's information. Provides direct access to Google Videos and Google Images
Facebook.com	World's largest social networking site that connects people
Amazon.com	Amazon.com is world's largest online retail company
Youtube.com	YouTube is world's largest collection of high quality videos
Yahoo.com	World's second largest internet search engine
Wikipedia.org	World's largest free encyclopedia available online which is built collaboratively by using wiki software
Ebay.com	World's leading person to person auction site, with products sorted into categories
Twitter.com	World's second largest social networking and microblogging site
Reddit.com	USA's largest collection of user-generated news links. Votes promote stories to the front page
Linkedin.com	India's largest professional networking site
Go.com	It is a searchable directory, news, stocks, sports and free e-mail
Craigslist.org	Craigslist provides local classifieds and forums for jobs, housing, for sale, personals, services, local community, and events
Imgur.com	It is used to share photos with social networks

	and on-line communities
Tumblr.com	Tumblr is a microblogging platform and social networking website that allows users to post multimedia and other content to a short-form blog
Espn.go.com	It is world's largest sports news network
Pinterest.com	Pinterest is an online place where users can post collections of things they love
Netflix.com	It is a flat monthly fee streaming TV and movies service
Live.com	Microsoft's internet search engine
Walmart.com	USA's leading retail company. It provides customers with on-line shopping experience of wide range of inventory items
Bing.com	Microsoft's internet search engine

Source: Alexa.com

Top 5 Websites in France

Website URL	Description
Google.fr	Version of Google in France
Facebook.com	World's largest social networking site that connects people
Google.com	World's largest internet search engine that allows its users to search for any of the world's information. Provides direct access to Google Videos and Google Images
Youtube.com	YouTube is world's largest collection of high quality videos
Amazon.fr	Version of Amazon.com in France

Source: Alexa.com

Top 25 Websites in the United Kingdom

Website URL	Description
Google.co.uk	UK version of Google.com
Facebook.com	World's largest social networking site that connects people
Google.com	World's largest internet search engine
Amazon.co.uk	UK version of Amazon.com, world's largest online retailer of books, movies, music and games along with electronics, toys, apparel, sports
Youtube.com	YouTube is world's largest collection of high quality videos
Ebay.co.uk	UK's person to person online auction site
Bbc.co.uk	The BBC Homepage
Yahoo.com	A major internet portal and service provider offering search results, and customizable content
Wikipedia.org	World's largest free encyclopedia built collaboratively by using wiki software
Live.com	Internet search engine from Microsoft
Twitter.com	World's leading social networking and microblogging site
Theladbible.com	The LAD Bible is one of the largest communities for guys aged 16-30 in the world
Linkedin.com	World's leading professional networking tool to find connections to recommended job candidates, industry experts and business
Dailymail.co.uk	UK's national tabloid offers news, sport, entertainment, and horoscopes
Paypal.com	Online payment service for individuals and merchants
Argos.co.uk	It offers appliances, DIY, electronics, furniture, garden supplies, gifts, jewellery, sports goods
Theguardian.com	Home of the Guardian, Observer and Guardian Weekly newspapers plus special-interest web sites
Tesco.com	UK's leading supermarket chain with online

	sales of groceries, clothing, and a range of goods for the home
Amazon.com	Amazon.com is world's largest online retail company
Msn.com	Microsoft Network Portal for shopping, news and money, e-mail, search, and chat
Bing.com	Search engine developed by Microsoft. Features web, image, video, local, news, and product search
Telegraph.co.uk	UK's leading newspaper that provides international online news
Bt.com	UK's national telecommunications provider
Imdb.com	Features plot summaries, reviews, cast lists, and theatre schedules
Ask.com	Offers search for web sites, images, news, blogs, video, maps and directions, local search

Source: Alexa.com

Top 10 Websites in Germany

Website URL	Description
Google.de	Google's German version
Facebook.com	World's largest social networking site
Amazon.de	Amazon's German version
Ebay.de	Ebay's German version
Youtube.com	YouTube is world's largest collection of high quality videos
Google.com	World's largest internet search engine
Wikipedia.org	World's largest free encyclopedia built collaboratively by using wiki software
Web.de	German Internet portal that provides free email service
Yahoo.com	A major internet portal and service provider offering search results, customizable content
Gmx.net	Provides free e-mail accounts with attachments up to 50 MB and POP3, IMAP and SMTP access

Top 10 Websites in Canada

Website URL	Description
Google.ca	Google's Canada website
Facebook.com	World's leading social networking site
Google.com	World's largest internet search engine
Youtube.com	YouTube is world's largest collection of high quality videos
Yahoo.com	A major internet portal and service provider offering search results, customizable content
Amazon.ca	The online shopping superstore in Canada
Wikipedia.org	World's largest free encyclopedia built collaboratively by using wiki software
Live.com	Search engine from Microsoft
Twitter.com	World's leading social networking and microblogging site
Amazon.com	World's largest online retail company

Source: Alexa.com

Top 20 Websites in Australia

Website URL	Description
Google.com.au	Australian version of Goolge.com
Google.com	World's largest search engine
Facebook.com	World's largest social networking site
Youtube.com	YouTube is world's largest collection of high quality videos
Yahoo.com	A major internet portal
Ebay.com.au	Australian version of ebay.com
Wikipedia.org	World's largest free encyclopedia
Linkedin.com	Professional networking tool to find connections
Amazon.com	Amazon.com is world's largest online retail company
Twitter.com	World's leading social networking and microblogging site
Live.com	Search engine from Microsoft
Bing.com	Search engine from Microsoft
Paypal.com	Online payment service for individuals and merchants
Gumtree.com.au	Site containing free local classified ads for flatshares, rentals, jobs, buy/sell, dating etc
News.com.au	News from Australia and the world, featuring national, world, business, sport, entertainment
Commbank.com.au	It offers personal banking, business solutions, institutional banking, company information, etc
Smh.com.au	The online edition of Sydney's Sunday newspaper with magazine sections on entertainment, health etc
Pinterest.com	Pinterest is an online pinboard where users can post collections of things they love
Realestate.com.au	Site containing real estate and property listings
Abc.net.au	The ABC is Australia's public broadcaster

Source: Alexa.com

TOP WEBSITES IN AGRIBUSINESS CATEGORY

Agriculture General
Top Websites in Agriculture Category

Website URL	Description
http://usda.gov/wps/portal/usda/usdahome	Homepage of United States Department of Agriculture (USDA)
http://www.nal.usda.gov/	Homepage of USDA National Agricultural Library
http://www.ars.usda.gov/main/main.htm	Homepage of USDA Agricultural Research Service
http://www.fao.org/index_en.htm	Homepage of Food and Agriculture Organization of the United Nations (FAO)
http://faostat.fao.org/	Homepage of FAOSTAT, the Statistics Division of the FAO
Scau.edu.cn	A key institution for higher agricultural learning
http://www.cgiar.org/	Homepage of CGIAR, a Global Agricultural Research Partnership
Cabdirect.org	An easily searchable archive of summaries of the world's agricultural and applied life sciences
Nrcs.usda.gov	Homepage of NRCS, USDA
Agricultureinformation.com	Worldwide agriculture directory
http://www.edis.ifas.ufl.edu/topics/agriculture/index.html	University of Florida IFAS Extension
Agriscape.com	An online directory on agriculture and its surrounding industry
Arborday.org	Information on tree and shrub care; education resources related to trees; promoting the planting
Ext.colostate.edu	Facts, tips, and directions for freezing vegetables
Lawnsite.com	The largest, original, and fastest growing forum online for the lawn and landscape industry

Extension.umn.edu	This site features instant access to the most recent Extension publications
Ces.ncsu.edu	Focuses its efforts on 20 major programs implemented by the county field faculty
Agr.gc.ca	Provides information, research, policies and programs to achieve security of the food system
Aphis.usda.gov	Information on animal and plant health/disease topics, USDA regulations, control of invasive species
Extension.psu.edu	Extension and outreach programs for producers, agricultural businesses and consumers
Groworganic.com	Supplies tools, equipment, and goods to organic gardeners and farmers
http://www.hort.purdue.edu/newcrop/med-aro/toc.html	Purdue University's Aromatic and Medicinal Plants Index
http://www.hort.purdue.edu/newcrop/afcm/	Purdue University's Alternative Field Crops Manual
http://www.hort.purdue.edu/newcrop/	Purdue University's the New Crop Resource Online Program
http://www.aginternetwork.org/en/	AGORA, Access to Global Online Research in Agriculture
http://www.agroatlas.ru/	AgroAtlas, Interactive Agricultural Ecological Atlas of Russia and Neighboring Countries
http://www.ipm.ucdavis.edu/	University of California Agriculture and Natural Resources: Integrated Pest Management Program
http://www.cipm.info/	NC State University's Center for Integrated Pest Management Program
http://www1.extension.um	University of Minnesota

n.edu/agriculture/	Extension
http://www.whatcom.wsu.edu/ag/compost/	Information on Composting from Washington State University
http://www.dpi.nsw.gov.au/agriculture	NSW Department of Primary Industries

Source: Alexa.com

Dryland Farming
Top Websites in Dryland Farming

Website URL	Description
Icrisat.org	With a mission to help developing countries in the semi-arid tropics increase farm productivity
Ialcworld.org	Information on sustainable agriculture for arid and semi arid regions
Drylandfarming.org	Provides information on a number of innovative ideas such as the use of medic pastures
Fao.org/docrep/U3160E/U3160E00.htm	On-line manual for the design and construction of water harvesting schemes
Syngentafoundation.org/index.cfm	Promotes improved and sustainable farming through better cultivation methods
Eden-foundation.org/project/aridland.html	Information from the Eden Foundation
Drylands-group.org	Network of various NGOs
Salineagriculture.com	SARC undertakes research into salt tolerant plants, especially cereals
Rala.is/rade/ralareport/Xinmin.pdf	A website about China's desertification. China is suffering from large scale and severe desertification

Source: alexa.com

Irrigation Systems
Top Websites in Irrigation Systems

Website URL	Description
Iwmi.cgiar.org	With a mission to contribute to food security and poverty eradication
Fao.org/nr/water/	Promotes efficient use and conservation of water resources to achieve food security
Irrigation.org	Updates on irrigation equipment
Twri.tamu.edu	Water news, subject index, keyword search, experts directory, research reports, about TWRI
Waterright.com.au	Provides information on water management and usage, simple irrigation systems
Esica.com	Manufactures irrigation instrumentation for monitoring and controlling water
Edis.ifas.ufl.edu/topic_irrigation	Covers a large number of topics including evapotranspiration, fertigation, irrigation practices etc
Cals.arizona.edu/crops/irrigation/azdrip/azdripindex.html	University of Arizona. Subsurface drip irrigation demonstration and research project
Ianrpubs.unl.edu/epublic/pages/index.jsp	University of Nebraska-Lincoln publications on feeding, breeding, herd management and business etc
Pods.dasnr.okstate.edu/docushare/dsweb/Get/Document-2222/e-951.pdf	Information on nursery management practices including water quality, best management, nutrition etc
Www3.telus.net/public/rolldr/indexdrip.htm	Calculations for designing trickle irrigation systems for trees and row crops
Irrigationtutorials.com/sprinkler00.htm	Describes what information to collect, what equipment one would need, and determine pipe size etc
Pivotirrigation.blogspot.com	Blog concerned with these irrigation systems and their components

Kulam.wordpress.com	Provides information on these water harvesting and storage systems found in most parts of South
Fertigasi.com/Mycrops/Fertigasi%20tech/tech.htm#What_is_Fertigation	What is fertigation?
Irrigation.org.au	An organisation representing the Australian irrigation industry
Itrc.org	University web site. Focusing on irrigation training and research
Wca-infonet.org	Publications, documents and data, computer programs, and discussion groups on the subjects
Automata-inc.com	Remote data acquisition for weather monitoring and irrigation control
Landandwater.com/features/vol44no3/vol44no3_1.html	Article describes a system comprised of a wetland and a water storage reservoir

Source: alexa.com

Precision Farming
Top Sites in Precision Farming

Website URL	Description
Navcen.uscg.gov	Provides the latest GPS and DGPS outage information, navigation forums
Agleader.com	This site showcases dedicated precision agriculture products
Farmworks.com	Offers a range of software for livestock management operations
Ayrstone.com	Provides an online forum and offers an educational DVD on GPS
Precisionplanting.com	Offers implements and training for precision farming
Omnistar.com	Worldwide differential global positioning systems with many widely-spaced base stations
Sstsoftware.com	Provides desktop and field software for precision agriculture applications
123farmworks.com	Provides precision farming software, equipment, and training on-site
Starpal.com	Mapping software for field use compatible with ESRI and MapInfo GIS, GPS mapping
Atsgps.net	Provides DGPS products and service, including brands such as Trimble, Ag Leader, TDS and StarPal
Cropmaps.com	Offers maps to show the spatial and temporal variability of fields
Gointime.com	Offers technology solutions to clients
Nzcpa.com	NZCPA is a precision agricultural department within Massey University
Rosenfeldsupply.com	Offers a connector protector to hold a GPS antenna cable in place
Rxvrt.com	Precision Variable Rate Programs and Precision Farming solutions using Precision Agriculture

Reduced Tillage Farming
Top Sites in Reduced Tillage Farming

Website URL	Description
No-tillfarmer.com	Providing the world's no-tillage farmers with ideas, inventions, techniques and industry news
Directseed.org	Information regarding direct deeding farming methods
Fao.org/docrep/007/ae371e/ae371e00.htm	Detailed FAO account of how soil erosion and other problems have been alleviated in Brazil
Pnwsteep.wsu.edu	Information on conservation tillage practices in the Pacific Northwest
Rolf-derpsch.com	Information about reduced tillage and no tillage practices, links to publications

Source: alexa.com

Agriculture Organizations
Top Sites in Agriculture Organizations Category

Website URL	Description
Fao.org	Site has many information resources and links on agricultural topics, as well as hunger, sustainable agriculture etc
Cgiar.org	With a mission to contribute to food security and poverty eradication in developing countries
Cirad.fr	Center for International Cooperation in Agronomy and Warm Climate Research
Cabi.org	Collects and shares knowledge about organic pesticides, biodiversity and genetics
Cta.int	International agency advancing agricultural and rural development in African, Caribbean, and Pacific
Ypard.net	Serves as a medium for Young Professionals (YPs) in Agricultural Research for Development (ARD)
Iica.int	Promotes the sustainable development of agriculture, food security and the prosperity of rural communities
E-agriculture.org	Global initiative to enhance sustainable agricultural development and food security
Globalgap.org	Non-governmental organization that sets voluntary standards for the certification of agricultural production systems
Agronomy.org	Dedicated to the development of agriculture enabled by science, in harmony with environment
Icid.org	News, profile and publications of the non-profit organisation working to improve water and land
Crops.org	An educational and scientific organization
Asabe.org	Professional and technical organization of members worldwide interested in engineering knowledge
Vitagora.com	European network composed of companies,

	research units and training organisations
Inbar.int	Knowledge dissemination for rural areas and industries on bamboo and rattan conservation and use
Slowfoodfounda tion.com	Promotes and funds projects to preserve "heritage" crops and livestock strains
Rangelands.org	For the scientific study, protection and management of rangeland resources, mainly grazing land
Alphazeta.org	A professional fraternity of men and women
Cambia.org	CAMBIA, a non profit biotech research organisation, giving a company profile, details of research
Apaari.org	Aims to promote the development of NARS in the Asia-Pacific region
Farmon.com	Farming, Farmers Markets, Farm Blogs, Farm Videos, Farming Artticles
Ict-agri.eu	Network of European national organizations aiming to coordinate research
Eaae.org	Information about the organisation, the European Review of Agricultural Economics, events, etc
Iaasworld.org	IAAS aims to promote the exchange of knowledge, information and ideas among students
Econbot.org	Fosters scientific research, education, and related activities on the past, present, and future

Source: alexa.com

Cooperative Extension Education
Top Sites in Cooperative Extension Education Category

Website URL	Description
Ext.colostate.edu	Facts, tips, and directions for freezing vegetables
Extension.umn.edu	This site features instant access to the most recent Extension publications, newsletters
Ces.ncsu.edu	Focuses its efforts on 20 major programs implemented by the county field faculty
Extension.psu.edu	Extension and outreach programs for producers, agricultural businesses and consumers
Www3.sdstate.edu	South Dakota State was founded as the state's only land-grant institution in 1881
Web.extension.illinois.edu	Provides practical, research-based information and programs to help individuals, families, farmers
Ext.vt.edu	Aims to improve the lives and well-being of people and communities in the state of Virginia
Extension.oregonstate.edu	Provides education and information based on timely research to help Oregonians solve problems
Extension.missouri.edu	Extends research and problem-solving resources from the University of Missouri
Clemson.edu/extension/	Provides extensive information related to home and gardening, farming, forestry, and emergency
Cce.cornell.edu	An educational system that

	enables people to improve their lives and communities
Msucares.com	Provides research-based educational programs and information in agriculture and natural resources
Aces.edu	The Alabama Cooperative Extension System
Uwex.edu/ces/index.cfm	Offers educational, research-based programs in agriculture, food and nutrition
Extension.usu.edu	Delivers research-based education and information to Utah residents in the areas of agriculture
Extension.uga.edu	Information for gardeners including pest management, vegetable growing
Uaex.edu	Offers research-based education programs and publications in the areas of agriculture
Extension.umd.edu	A statewide, non-formal education system within the university
Ksre.ksu.edu	Offers educational programs and resources in the areas of agribusiness and economics
Csrees.usda.gov	Advanced knowledge for agriculture, the environment, human health and well-being
Extension.arizona.edu	Serves as a statewide network of faculty and staff that provides lifelong educational programs
Ucanr.org	Offers educational programs throughout California in the areas of farm management
Extension.unh.edu	Find resources covering

	agriculture, forestry, family, 4-H, youth, Sea Grant, water
Solutionsforyourlife.ufl.edu	Extension is a partnership between state, federal, and county governments
Unce.unr.edu	Delivers non-degree, educational programs in these areas of agriculture; children, youth

Source: alexa.com

Agriculture Directories
Top Sites in Agriculture Directories

Website URL	Description
Agricultureinformation.com	Worldwide agriculture directory
Agriscape.com	An online directory on agriculture and its surrounding industry
Hort.purdue.edu/newcrop/	From Purdue University
Prairielinks.com	Canadian directory providing links, including farm equipment, agribusiness, livestock
Intute.ac.uk/agriculture/	Evaluated web resources aimed at students, researchers, academics and practitioners
Berrygrape.org	An information and communications resource for berry and grape production
Ec.europa.eu/research/agriculture/scar/index_en.cfm	Information about the various parts of the programme and related activities
Agry.purdue.edu/links/	Provided by Purdue University
Cipm.ncsu.edu/agVL/	Extensive collection of resources and links

Library.wur.nl/way/search/units.html	Scientific publications from current and former research units in the fields of food

Agriculture Databases
Top Websites in Databases

Website URL	Description
Orton.catie.ac.cr	A collection of agricultural libraries for Latin America and the Caribbean
Faostat.fao.org	Online and multilingual databases currently containing over 1 million time-series records
Fao.org/countryprofiles	The system integrates country-based information from FAO's thematic and statistical databases
Agricola.nal.usda.gov	Bibliographic database of citations to the agricultural literature created by the National Agricultural Library
Nasdonline.org	Database of materials devoted to increased safety, health and injury prevention in agriculture
Aginternetwork.org/en/	Allows for searching journal titles and with login, full text articles
Theagricos.com	Provides information on topics related to agriculture, biotechnology, plant breeding, genetics
Fisaonline.de	Provides an overview of publicly funded research projects in the agricultural and food science
Amis-outlook.org	The global agricultural market information system
Seedimages.com	Database with information on seed identification, dormancy, testing and analysis, development
Pestmanagement.info/nass/	National Agricultural Statistics Service portal offers search for statistics by year, state etc

Archives.gov/research/formats/electronic-records.html	The National Archives and Records Administration provides access to statistical reports
Cris.csrees.usda.gov	USDA funded portal offers descriptions of current and completed research projects
Fas.usda.gov/psdonline/psdHome.aspx	Provides current and historical official data on production, supply and distribution
Fao.org/corp/topics/en/	Provides links to FAO information on a wide range of topics including desertification

Source: alexa.com

Apiculture
Top Sites in Apiculture-Honey Bees Category

Website URL	Description
Honey.com	Information for consumers, the food industry and the honey industry
Beesource.com/forums/	Several forums for discussion of general beekeeping topics
Apiservices.com	International site with beekeeping information available in several different languages
Ocbeekeepers.org	Formed in the 1970's we are the oldest and largest beekeeping organization in Orange County, CA
Buzzaboutbees.net	Describes the different types of bees, their nests, life cycles, causes of decline, honey, etc
Beehoo.com	World-wide beekeeping directory
Beethinking.com	Supplier of horizontal top bar and Warré hives
Ibra.org.uk	A not-for-profit organisation, based in Cardiff
Biobees.com	Beekeeping using top bar hives
Coloss.org	The international, non-profit association is focussed on improving the well-being of bees
Beeculture.com	The magazine of American beekeeping, with current and archive articles available on-line
Beemaster.com	An international beekeeping forum
Beverlybees.com	Organic, Treatment Free Beekeeper who provides videos, how-to's and beekeeping instructional program
Abfnet.org	Acts on behalf of the beekeeping industry
Beeinformed.org	An extension project trying to decrease winter mortality of managed honey bee colonies
Ontariobee.com	Established in 1881 in Ontario,

	Canada
Honeylove.org	A nonprofit conservation organization with a mission to protect the honeybees
Easternapiculture.org	"Educating the Beekeeping Community" principally through a 5-day annual conference
Honeybee.com.au	Includes beekeeping information, image gallery, chat page, Bee facts, services, honey recipes etc
Bushfarms.com/bees.htm	Natural beekeeping without treatments for pests or diseases and only minimal interventions
Bees-on-the-net.com	Practical information on all aspects of the honeybee and how to start beekeeping
Jordanbru.info	Dr. Nizar Haddad on beekeeping in Jordan - books, articles and films about beekeeping
Americanbeejournal.com	Specialist publication available by subscription including beekeeping information, education etc
Ncbeekeepers.org	Provides information on the Master Beekeeper Program, news, events, membership benefits etc
Kiwimana.co.nz	Organic Beekeepers from the hills of West Auckland in New Zealand

Source: alexa.com

TOP WEBSITES IN HORTICULTURE CATEGORY

Horticulture General
Top Websites in Horticulture

Website URL	Description
Lawnsite.com	The largest, original, and fastest growing forum online for the lawn and landscape industry
Aggie-horticulture.tamu.edu	Horticulture feature and topic areas, extension resources, and student and program information
Actahort.org	Online books and articles of the magazine "Acta Horticulturae"
En.wikipedia.org/wiki/Banana	Information from Wikipedia on this fruit, including its description, world trade, cultivation etc
Mushroominfo.com	Large collection of mushroom recipes. Pick recipes by variety, food course, or cuisine
Hortsci.ashspublications.org	On-line publication by the American Society for Horticultural Science, with seven issues a year
Carrotmuseum.co.uk	A virtual museum whose mission is to educate, inform and amuse visitors through the collection
Gcsaa.org	Golf course management information, member areas, and a research section which covers requests
Avocadosource.com	News, information, links and chat for avocado growers in many of the leading growing nations
Carbon.org	Site covers much of the wild and domesticated flora of utility to humans, and their ecosystems etc
Gopherforum.com	Lawn care business online discussion community
Internationaloliveoil.org	Contains the group's aims, activities, international agreements, olive oil and table olive status etc
Strawberryplants.or	The one stop for everything related to

g	strawberry plants and growing strawberries
Gernot-katzers-spice-pages.com/engl/	Reference information about more than 100 herbs and spices, plus their usage in ethnic cuisines
Horttech.ashspublications.org	Produced four times per year by the American Society for Horticultural Science
Gardenorganic.org.uk	Henry Doubleday Research Association; Advice and Heritage Seed Library
Worldcocoafoundation.org	Promotes a sustainable cocoa economy through economic and social development
Ishs.org	Includes general and specific scientific information, links to journals, events calendar
Wineserver.ucdavis.edu	Academic study of winemaking from the University of California, Davis in the USA
Bananalink.org.uk	Details of campaigns, lobbies, and researches on social and environmental issues concerned with banana
Journal.ashspublications.org	Provides access to abstracts, full texts and PDF versions of articles in this bi-monthly magazine
Bananas.org	Homepage of the International Banana Society
Dpi.nsw.gov.au/agriculture/horticulture	Australian government agency provides information on home and commercial horticulture of fruits
Ashs.org	Promotes and encourages scientific research and education in horticulture within the United States
Crec.ifas.ufl.edu	University of Florida research, teaching and extension center devoted to citrus

Source: Alexa.com

Horticulture Publications
Top Websites in Horticulture Publications Category

Website URL	Description
Actahort.org	Online books and articles of the magazine "Acta Horticulturae"
Hortsci.ashspublications.org	On-line publication by The American Society for Horticultural Science, with seven issues a year
Horttech.ashspublications.org	Produced four times per year by the American Society for Horticultural Science
Journal.ashspublications.org	Provides access to abstracts, full texts and PDF versions of articles
Growingmagazine.com	Online magazine published by Moose River Media covering fruit, nut and vegetable production
Horticultureworld.net	Information about the horticultural technology with links to the Journal of Applied Horticulture
Gmpromagazine.com	Online magazine for growers with news, industry information, multimedia, community and resources
Jpacd.org	Articles cover scientific and industry news concerning growing cacti
Horticultureresearch.net	International journal published twice a year
Hortsci.ashspublications.org/archive	Abstracts of articles containing original research and papers presented at workshop

Source: Alexa.com

Ornamentals and Floriculture
Top Websites in Ornamentals and Floriculture Category

Website URL	Description
Extension.umass.edu/floriculture/	University of Massachusetts resource, with information on educational programs and grower services
Hort.ifas.ufl.edu/woody/	Account of the production of ornamentals in Florida
Ces.ncsu.edu/depts/hort/floriculture/	The starting page for access to North Carolina State University's floriculture information
Endowment.org	Funds research and educational development addressing floral industry needs
Usna.usda.gov/Research	The unit conducts a broad based program contributing to basic and developmental research
Edis.ifas.ufl.edu/IG012	Information on deciding how and when to treat pests on ornamentals as part of an integrated application
Edis.ifas.ufl.edu/IG125	Information on the biology and life cycle of pests on ornamentals, and their management strategies
Hort.cornell.edu/department/faculty/wmiller/bulb/index.html	Describes the bulb research program at Cornell University
Ohioline.osu.edu/hyg-fact/1000/1253.html	A fact sheet from Ohio State University
Edis.ifas.ufl.edu/EP060	Information on growing and propagating this bulbous plant, and the diseases and pests that may affect the crop
Edis.ifas.ufl.edu/IN398#TABLE_15	Details of suitable pesticides to use to manage these pests, provided by the University of Florida
Edis.ifas.ufl.edu/IG144	IPM involves careful use of pesticides in coordination with other pest management practices
Edis.ifas.ufl.edu/IN715	Information on the development of

	resistance to insecticides and miticides and how to deal with them
Edis.ifas.ufl.edu/IG110	Discusses the problems involved in managing pests in peopled areas
Edis.ifas.ufl.edu/PP164	Illustrated guide to the symptoms of this condition, its diagnosis and management
Umass.edu/umext/floric ulture/fact_sheets/green house_management/pgr .html	Describes the reasons, effects and usage of these products
Entoplp.okstate.edu/dd d/insects/aphids.htm	Photograph of an infected leaf with information on hosts, symptoms, life cycle etc
Ag.auburn.edu/hort/lan dscape/Elily.htm	Detailed description on commercial bulb and flower raising for the cut flower trade
Ag.auburn.edu/hort/lan dscape/Hcactus.htm	Provides instructions for the commercial greenhouse production of Schlumbergera bridgesii
Cabrillo.edu/academics/ horticulture/salvias/htm l	Information about the college's stock of cultivated salvia, list of cultivars etc
Members.tripod.com/~ Hatch_L/nos.html	Devoted to the study of new and rare ornamental landscape plants
Members.tripod.com/~ Hatch_L/genera.html	Online plant finder covering the United States, lists of nurseries, descriptions of plants
Pinoyhorticulture.blogsp ot.com	Blog of activities of horticulture societies and plant enthusiasts in the Philippines
Pinoyhorticulture.wordp ress.com	Provides information on the activities of horticulture societies and plant enthusiasts
Delworth.ca	Non-profit group promoting research and education in the Canadian floriculture industry

Mushrooms and Edible Fungi
Top Websites in Mushrooms and Edible Fungi Category

Website URL	Description
Mushroominfo.com	Large collection of mushroom recipes
Mushroomcompany.com	News of the mushroom industry, marketing opportunities, events, recipes, growing information etc
Plantationsystems.com	Undertakes research on black truffles (Tuber melanosporum)and operates a global network of clients
Oystermushrooms.net	Includes an online book that provides detailed advice on growing oyster mushrooms
Americanmushroom.org	Trade association representing growers, processors, and marketers of cultivated mushrooms
Mushroomspawn.cas.psu.edu/mushroom.shtml	Information on the cultivation and production of the common mushroom from Penn State University
Botit.botany.wisc.edu/toms_fungi/apr2001.html	Description of the available strains, and cultivation methods
Ucce.ucdavis.edu/datastore/detailreport.cfm	The official site of UC Davis
Ars.usda.gov/research/projects/projects.htm	The main in-house research arm of the U.S. Department of Agriculture
Nfs.unl.edu/documents/SpecialtyForest/Schoepski.pdf	Discussion of this most important part of mushroom production, including distribution, packaging etc
Ca.uky.edu/agc/pubs/for/for77/for77.pdf	A pictorial guide from the University of Kentucky cooperative extension
Extension.missouri.edu/explorepdf/agguides/agroforestry/af1010.pdf	Well illustrated account of the processes involved in Shiitake production
Courses.wcupa.edu/j	Mushrooms are the number one cash

ones/his480/reports /mushroom.htm	crop in the state of Pennsylvania
Mykoweb.com/articl es/gardendemo.html	Provides information on the designing and setup of a year-round mushroom demonstration garden
Attra.org/attra- pub/mushroom.html	Detailed notes on how to set up a production system for mushroom production

Source: Alexa.com

Shiitake Mushrooms
Top Websites in Shiitake Mushrooms Category

Website URL	Description
Ars.usda.gov/rese arch/projects/pro jects.htm	The main in-house research arm of the U.S. Department of Agriculture
Nfs.unl.edu/docu ments/SpecialtyF orest/Schoepski.p df	Discussion of this most important part of mushroom production, including distribution, packaging etc
Ca.uky.edu/agc/p ubs/for/for77/fo r77.pdf	A pictorial guide from the University of Kentucky cooperative extension
Extension.missour i.edu/explorepdf/ agguides/agrofore stry/af1010.pdf	Well illustrated account of the processes involved in Shiitake production
Mushroomcompa ny.com/resources /shiitake/shiitake. shtml	The Mushroom Growers' Newsletter provides an introduction to growing shiitake

Source: Alexa.com

Herbs
Top Websites in Herbs Category

Website URL	Description
Hort.purdue.edu/newcrop/CropFactSheets/basil.html	Factsheet on the uses and cultivation of basil
Hort.purdue.edu/newcrop/afcm/comfrey.html	Factsheet on comfrey, its history, uses, varieties and cultivation
Hort.purdue.edu/newcrop/med-aro/factsheets/HORSERADISH.html	Factsheet about horse radish
Hort.purdue.edu/newcrop/afcm/mustard.html	Factsheet on mustard, its history, uses, varieties and cultivation
Hort.purdue.edu/newcrop/SavoryHerbs/SavoryHerbs.html	Information on a number of herbs and how to grow them

Source: Alexa.com

Lavender
Top Sites in Lavender Category

Website URL	Description
Pss.uvm.edu/pss123/herlaven.html	Basic description, botanical characteristics, and list of principal cultivars
Beyond.fr/flora/lavender.html	Describes the lavenders native to the Southern French Alps

Source: alexa.com

Hydroponics
Top Websites in Hydroponics Category

Website URL	Description
Growingedge.com	Information on hydroponics, aquaponics and greenhouses for commercial and hobby growers
Hydroponiceconomics.com /blog/	A blog focused on hydroponics and building efficient home hydroponic systems
Ag.arizona.edu/hydroponictomatoes/	Interactive web site from the University of Arizona provides practical, accurate information
Hydroponic-gardens.com	A site loaded with tips and step by step instructions for building your own hydroponic systems
Hydroponics-center.com	Offers information on hydroponics systems and growing techniques
Mayhillpress.com	Secrets for starting hydroponic businesses
Hydroponicsonline.com/lessons/table-of-contents.htm	Provides a course on hydroponic cultivation
Schundler.com/hydroculture.htm	A research report issued by the Perlite Institute, Inc. by Dr. David A. Hall
Cocoponics.co	Technical blog sharing the author's experiences of hydroponics

Source: Alexa.com

Propagation and Nursery Practice
Top Websites in Propagation and Nursery Practice Category

Website URL	Description
Oregonstate.edu/dept/ nursery-weeds/	Discussions on weed control, weed identification, and herbicide management in nursery crops
Canr.org	A collaborative effort between the University of Georgia and McCorkle Nurseries, Inc.
Rooting.ucdavis.edu/P CHOME.HTM	Provides propagation information, including required chemicals and timetables etc
Edis.ifas.ufl.edu/TOPI C_Nursery_Propagatio n	Collection of articles on plant production, equipment, and nursery practices, with illustration
Aggie- horticulture.tamu.edu/ greenhouse/nursery/g uides/	Provides information on topics such as different types of structures, heating requirements etc

Source: Alexa.com

Spices
Top Websites in Spices Category

Website URL	Description
Gernot-katzers-spice-pages.com/engl/	Reference information about more than 100 herbs and spices, plus their usage in ethnic cuisines
Theepicentre.com/spice/white-turmeric-zedoary/	General information on this spice, its preparation and storage, culinary and medicinal uses
Theepicentre.com/spice/paprika/	General information on this spice, its preparation and storage, culinary and medicinal uses
Fao.org/docrep/v5350e/V5350e04.htm	Provides information on the oils extracted from the bark of trees of the genus Cinnamomum
Nhm.ac.uk/jdsml/nature-online/seeds-of-trade/print.dsml	Introduction to the history of cultivation and spread of some common crops
Kew.org/plant-cultures/plants/black_pepper_production__trade.html	Information on the international trade in pepper, its cultivation and harvest, and the processing
Kew.org/plant-cultures/plants/chilli_pepper_production__trade.html	Information on the cultivation, marketing and processing of chilli pepper in South Asia
Kew.org/plant-cultures/plants/tamarind_production__trade.html	Information on cultivating and harvesting tamarind crop
Kew.org/plant-cultures/plants/turmeric_production__trade.html	Information on cultivating and harvesting turmeric crop
Gernot-katzers-spice-pages.com/engl/Croc_sat.html	Information about saffron, its cultivation, names and uses. Includes photos

Naturland.de/filead min/MDB/docume nts/Publication/En glish/pepper.pdf	Covers all aspects of pepper plant cultivation, biological methods of plant protection, harvesting etc
Naturland.de/filead min/MDB/docume nts/Publication/En glish/vanilla.PDF	Covers all aspects of vanilla plant cultivation, biological methods of plant protection, harvesting etc
Theepicentre.com/s pice/cassia/	General information on cassia, its preparation and storage, culinary and medicinal uses
Theepicentre.com/s pice/cloves/	Information on cloves, its history, spice description, preparation and storage, uses etc
Theepicentre.com/s pice/ginger/	Information on ginger, its history, spice description, preparation and storage, uses etc
Theepicentre.com/ growing-your-own- ginger/	Cultivation of ginger, Zingiber officinale
Theepicentre.com/s pice/mustard/	General information on mustard, its preparation and storage, culinary and medicinal uses etc
Theepicentre.com/s pice/nutmeg/	Information on nutmeg, its history, spice description, preparation and storage, uses, etc
Theepicentre.com/s pice/turmeric/	General information on turmeric, its preparation and storage, culinary and medicinal uses etc
Https://rirdc.infose rvices.com.au/dow nloads/00-155.pdf	Report from the Australian Rural Industries Research and Development Corporation

Turfgrass
Top Websites in Turfgrass Category

Website URL	Description
Lawnsite.com	The largest, original, and fastest growing forum online for the lawn and landscape industry
Aggie-horticulture.tamu.edu	Horticulture feature and topic areas, extension resources, and student and program information
Gcsaa.org	Golf course management information, member areas, and a research section which covers requests
Gopherforum.com	Lawn care business online discussion community
Jcgolf.com	International golf and sports turf consulting company focusing on agronomic, environmental issues
Lawnchat.com	Discussions on running a lawn care business, caring for turfgrass and maintaining lawn care
Turffiles.ncsu.edu	Information from North Carolina State University professionals, including weed identification
Flrec.ifas.ufl.edu	Turfgrass, urban entomology, aquatic plant management and environmental horticulture
Ipm.ucdavis.edu/tools/turf/turfspecies/	Introduces the grass species used for turf in California
Plantfacts.ohio-state.edu	Collection of horticultural resources, including a search engine drawing from U.S. university

Source: alexa.com

Viticulture
Top Websites in Viticulture Category

Website URL	Description
Wineserver.ucdavis.edu	Academic study of winemaking from the University of California, Davis in the US
Winegrowers.info	Provides information on vine varieties, clones and rootstocks, planting a vineyard, wine making
Awri.com.au	Provider of research and development services to the Australian wine industry
Vineyardteam.org	Promotes environmentally and economically sustainable vineyard practices on the Central California
Grapegrowingguide.com	Provides information on starting a vineyard, grape cultivation, making home-made wine
Asev.org	Information on the society, research and online copies of the society's journal
Winegrapes.tamu.edu	Viticulture information resource for wine grape growers
Brocku.ca/ccovi/	The site features information about the Institute, Faculty research activities, academic information
Concordgrape.org	Represents processors of Concords and manufacturers of products derived from Concords
Grapeseek.org	The grapes and wine search engine. Drive traffic to your

	wine industry website
Avf.org	Nonprofit supporting research in viticulture and wine making
Liveinc.org	A sustainable agriculture program providing vineyards and wineries with official recognition
Grapes.umn.edu	Development of hardy, disease resistant, quality wine grape varieties etc
Nysaes.cornell.edu/fst/faculty/henick/	Evaluations of grape cultivars, grape growing and vinification techniques on wine quality
Edis.ifas.ufl.edu/IG071	Information on insects which attack this crop, with details of insecticides registered for use
Edis.ifas.ufl.edu/AG208	Information on the achievements made by scientists at the Florida Agricultural Experiment Station
Dpi.nsw.gov.au/research/areas/production-research/viticulture	Information on the research program at the National Wine and Grape Industry Centre in Australia
Viticulture.hort.iastate.edu	Articles and bulletins relating to general viticulture
Wine.wsu.edu/virology/	Comprehensive information on wine grape virus diseases including leafroll disease
Academic.sun.ac.za/agric/departments/dww-iwbt_eng.html	Extensive information on university courses and applications of viticulture in South Africa
Csu.edu.au/nwgic/	This Australian organisation aims to assist the national

	industry by means of research, education etc
Lincoln.ac.nz/Research-at-Lincoln/Research-centres/Centre-for-Viticulture-and-Oenology/	Provides details of the wine research being undertaken at Lincoln University in Canterbury
Jisvv.com	Publishes original peer reviewed research reports, short research notes and review papers
Berrygrape.org	An information and communications resource for berry and grape production
Ontariograpes.com	A group of commercial wine grape growers and wineries, bringing viticulture and enology together

Source: alexa.com

Fruits
Top Websites in Fruits Category

Website URL	Description
En.wikipedia.org/wiki/Banana	Information from Wikipedia on this fruit, including its description, world trade, cultivation etc
Avocadosource.com	News, information, links and chat for avocado growers in many of the leading growing nations
Internationaloliveoil.org	Contains the group's aims, activities, international agreements, olive oil etc
Strawberryplants.org	The one stop for everything related to strawberry plants and growing strawberries
Worldcocoafoundation.org	Promotes a sustainable cocoa economy through economic and social development
Bananalink.org.uk	Details of campaigns, lobbies, and researches on social and environmental issues
Bananas.org	Homepage of the International Banana Society
Crec.ifas.ufl.edu	University of Florida research, teaching and extension center devoted to citrus
Hort.purdue.edu/newcrop/morton/	List and description of fruits, including lesser-known varieties, that grow in tropical and subtropical regions
Ultimatecitrus.com	Information and links regarding anything citrus related, from agriculture to weather
Blueberry.org	Promoting highbush blueberry growing. Information on the plant, growth, and the product
Pnwhandbooks.org/plantdisease/	Guide to the control and management tactics for the more important plant diseases in the Pacific
Dpi.nsw.gov.au/agriculture/horticulture/pom	Information on these fruit including fertilizers, rootstocks, varieties, crab

es	apples
Dpi.nsw.gov.au/agriculture/horticulture/citrus	Collection of articles by Australian governmental agency on cultivating citrus, pests and diseases
En.wikipedia.org/wiki/Black_sigatoka	Information from Wikipedia on this leaf spot disease of bananas
Dpi.nsw.gov.au/agriculture/horticulture/pomes/quince-growing	Factsheet discussing cultivation, varieties, diseases, and harvesting of quinces
Almondboard.com	Providing up to date information about the almond industry
Extension.psu.edu/plants/gardening/fphg	A number of tree fruit and other resources from the Penn State University Horticulture Department
Strawberryplants.org/2010/05/growing-strawberries/	Comprehensive guide for the home gardener who wants to grow strawberries
Virginiafruit.ento.vt.edu	Fact sheets on the cultivation of various fruit crops
Cranberryinstitute.org	Conducts research into the crop. Includes brief details of recent work and findings
Indiaagronet.com/indiaagronet/crop%20info/strawberry.htm	A leading Indian website dealing with strawberry information
Cocoainitiative.org	Partnership between NGOs, labour unions, cocoa processors and the major chocolate brands
Wvu.edu/~agexten/hortcult/fruits/prunblkbr.htm	Information on the cultural techniques for fruits
Nutgrowing.org/links.htm	Links to resources on nut tree growing and nut tree cultivars

Source: alexa.com

Almond
Top Websites in Almond Category

Website URL	Description
Almondboard.com	Providing up to date information about the almond industry
Homeorchard.ucdavis.edu/plant_almond.pdf	Information on five varieties commonly grown in California
Homeorchard.ucdavis.edu/8126.pdf	Information on reducing the potential for on-farm contamination of almonds
Californiaagriculture.ucanr.org/landingpage.cfm	A bimonthly magazine of news and peer-reviewed research
Ipmcenters.org/CropProfiles/docs/caalmonds.pdf	Besides information on general production and cultural practices

Source: alexa.com

Apple
Top Websites in Apple Category

Website URL	Description
Dpi.nsw.gov.au/agriculture/horticulture/pomes	Information on these fruit including fertilizers, rootstocks, varieties etc
Orchard.uvm.edu	Extension and Research for the commercial tree fruit grower in Vermont and beyond
Ohioline.osu.edu/hyg-fact/1000/1150.html	Article on how and when to prune these fruit trees
Devon-apples.co.uk	An academic resource about apples and growing them in Devon, recording the many apple varieties
Ohioline.osu.edu/hyg-fact/1000/1403.html	Provides information on these physiological disorders that can affect apple quality
Okspecialtyfruits.com	This Canadian agricultural biotechnology company is developing new commercial tree fruit variety
Ohioline.osu.edu/hyg-fact/1000/1401.html	Article outlining what cultivars to choose, the size of trees, choosing a site, planting, pruning etc
Dpi.nsw.gov.au/__data/assets/pdf_file/0004/154669/zinc-deficiency-in-apples.pdf	Information on the causes of this condition, the symptoms, prevention and treatment
Agroatlas.ru/cultural/Malus_domestica_K_en.htm	Information on the apple tree, its biology and morphology, distribution in the former USSR
Ipmcenters.org/Crop Profiles/docs/NewEnglandApples.pdf	General production information on this crop, the cultural practices involved in growing apples

Source: alexa.com

Banana
Top Websites in Banana Category

Website URL	Description
En.wikipedia.org/wiki/Banana	Information from Wikipedia on this fruit, including its description, world trade, cultivation etc
Bananalink.org.uk	Details of campaigns, lobbies, and researches on social and environmental issues
Bananas.org	Homepage of the International Banana Society
En.wikipedia.org/wiki/Black_sigatoka	Information from Wikipedia on this leaf spot disease of bananas
Crfg.org/pubs/ff/banana.html	Factsheet on the cultivation of this crop with special reference to California
En.wikipedia.org/wiki/List_of_banana_and_plantain_diseases	From Wikipedia, a list of the bacterial, fungal and virus diseases and parasitic nematodes
Hort.purdue.edu/newcrop/morton/banana.html	Describes the origin, growth, cultivars, cultivation, and uses of the eating banana
Hort.purdue.edu/newcrop/faminefoods/ff_families/MUSACEAE.html	Lists some of the food use of less familiar members of the Musaceae
Edis.ifas.ufl.edu/IN706	Illustrated information on this pest, its distribution, description, life cycle, biology, hosts etc
Aphis.usda.gov/publications/plant_health/content/printable_version/pa_rpm7-2007.pdf	Information from USDA on this invasive pest of coconut palms and banana plants

Source: alexa.com

Cane Fruits
Top Websites in Cane Fruits Category

Website URL	Description
Wvu.edu/~agexten/ hortcult/fruits/prunb lkbr.htm	Information on cultural techniques for cane fruits
Extension.umaine.ed u/publications/2066e /	Details on growing these cane fruits, including site selection, training and pruning
Gardenaction.co.uk/f ruit_veg_diary/fruit_ veg_mini_project_jan uary_2_raspberry.asp	Information on the cultivation of this fruit, the planting, support and care of canes etc
Www1.agric.gov.ab.c a/$department/deptd ocs.nsf/all/faq6860	Information on the timing of herbicide applications and the products that are suitable for these applications
Ipmcenters.org/Crop Profiles/docs/cacane berries.pdf	General production information on raspberries and blackberries, production practices etc
Gardenaction.co.uk/f ruit_veg_diary/black berry_page1.asp	Information on the cultivation of this fruit, the planting and care of canes, choice of varieties etc
Wvu.edu/~agexten/ hortcult/fruits/bram bles.htm	Information on soils, site preparation, planting, pruning and managing raspberries and blackberries etc
Assuredproduce.co.u k/resources/000/255 /624/Fruit_(Cane_Fr uit)1.pdf	Producing crops in accordance with these standards allows growers to sell produce
Farminfo.org/orchar d/blackberry.htm	Information on cultivating these fruit including planting, varieties and harvesting etc
Farminfo.org/orchar d/raspberry.htm	Plenty of information on cultivating this crop, recommended varieties, planting and growing etc

Source: alexa.com

Citrus Fruits
Top Websites in Citrus Fruits Category

Website URL	Description
Crec.ifas.ufl.edu	University of Florida research, teaching and extension center devoted to citrus, addressing issues
Ultimatecitrus.com	Information and links regarding anything citrus related, from agriculture to weather
Dpi.nsw.gov.au/agriculture/horticulture/citrus	Collection of articles by Australian governmental agency on cultivating citrus, pests and diseases etc
Users.kymp.net/citruspages/	Description of over 330 types of citrus
Edis.ifas.ufl.edu/TOPIC_Citrus	Species descriptions and cultivation practices
Edis.ifas.ufl.edu/CG002	Information on mites that attack citrus, the application of miticides and chemical control
Edis.ifas.ufl.edu/HS318	Article with photographs on the vine weeds that can be found in Florida citrus groves
Aggie-horticulture.tamu.edu/citrus/tamuhort.html	Information and resources for citrus growers
Dpi.nsw.gov.au/__data/assets/pdf_file/0009/137682/dwarfing-citrus-trees-using-viroids.pdf	Information on research in Australia into using graft-transmissible dwarfing budlines to control
Fao.org/docrep/T0601E/T0601E00.htm	Information on techniques and laboratory methods for the biological detection

Source: alexa.com

Currants and Gooseberries
Top Websites in Currants and Gooseberries Category

Website URL	Description
Ars-grin.gov/cor/ribes/ribsymp/ribsymp.html	Detailed illustrated information on disease, insect and other problems affecting these fruit
Gardenaction.co.uk/fruit_veg_diary/blackcurrant_page1.asp	Information on growing this fruit, the choice of varieties and planting and care of bushes etc
Currants.com/index.php	Information about Currants

Fig
Top Websites in Fig Category

Website URL	Description
Hort.purdue.edu/newcrop/morton/fig.html	A detailed account of the tree's uses, origin, botany and cultivation
Caes.uga.edu/Publications/displayHTML.cfm	Agricultural and applied economics department web site with information for prospective students
Ipmcenters.org/CropProfiles/docs/cafigs.pdf	General information on commercial fig production in this state, its cultural requirements etc

Source: alexa.com

Mango
Top Websites in Mango Category

Website URL	Description
Hort.purdue.edu/newcrop/morton/mango_ars.html	A chapter covering history of cultivation, botanical description, comparison of varieties etc
Fao.org/fileadmin/user_upload/inpho/docs/Post_Harvest_Compendium_-_Mango.pdf	Provides information on harvesting, packinghouse operations, cooling system, storage, transport etc
Kew.org/plant-cultures/plants/mango_production__trade.html	Information on the international trade in mangos, the cultivation of the trees and processing
Wwwchem.uwimona.edu.jm:1104/lectures/mango.html	Includes list of Jamaican varieties and a chemical analysis of fruit
Naturland.de/fileadmin/MDB/documents/Publication/English/mango.pdf	Covers all aspects of plant cultivation, harvesting, processing, handling, packing and storage

Source: alexa.com

Nuts
Top Websites in Nuts Category

Website URL	Description
Nutgrowing.org/links. htm	Links to resources on nut tree growing and nut tree cultivars
Songonline.ca/nuts/de fault.htm	Information on several different types of nut tree suitable for growing in Ontario
Acf.org/pdfs/about/re storation.pdf	Information on the breeding of chestnut trees resistant to chestnut blight
Tech.groups.yahoo.co m/group/CICLY/	Centre for Information on Coconut Lethal Yellowing is intended to act as a clearinghouse
Hort.purdue.edu/newc rop/NewCropsNews/ 94-4-1/nuts.html	Information on site requirements and suitable cultivars for the cultivation of filberts, chestnut etc
Edis.ifas.ufl.edu/IG07 7	Information on the timing, coverage and rate of use of sprays to control pests in this crop
Pecankernel.tamu.edu/ introduction/	Provides an overview of the pecan tree, and information on the pests and diseases
Caes.uga.edu/Publicati ons/pubDetail.cfm	Agricultural and applied economics department web site with information for prospective students
Fao.org/docrep/005/a c451e/ac451e03.htm	An account by the FAO of the status of this crop in China and the production methods used there
Extension.missouri.edu /xplor/miscpubs/mp0 711.htm	Photographs and information on the problems that may be encountered in growing this crop
Sres-associated.anu.edu.au/f pt/nwfp/macanut/ma canut.html	Information on these nuts, their history, distribution, cultivation, nutritional benefits etc
Kew.org/plant-cultures/plants/cocon	Information on the coconut, its cultivation, harvesting and processing,

ut_production__trade.html	and the uses etc
Nybg.org/bsci/braznut/	Article by Scott A. Mori on Bertholletia excelsa, its natural history, the harvesting of the nuts etc
Attra.org/attra-pub/pecan.html	Information on the pecan, Carya illinoinensis, its cultivation including sustainable and organic practices
Msucares.com/pubs/infosheets/is0493.pdf	Factsheet from the Mississippi State University Extension
Naturland.de/fileadmin/MDB/documents/Publication/English/coco_palm.pdf	Covers all aspects of coconut cultivation, harvesting, processing, handling, packing and storage
Naturland.de/fileadmin/MDB/documents/Publication/English/macadamia.pdf	Covers all aspects of plant cultivation, biological methods of plant protection, harvesting etc
Naturland.de/fileadmin/MDB/documents/Publication/English/brazil_nut.pdf	Covers Brazil nut production, harvesting, storing, processing and packing
Amazonconservation.org/ourwork/livelihoods.html	The Brazil Nut Project, initiated by the Amazon Conservation Association
Oregonhazelnuts.org/research.php#micro	Information from the Oregon Hazelnut Marketing Board on their research including their hazelnuts
Ipmcenters.org/CropProfiles/docs/capistachios.pdf	General production information on this crop, the insect pests and diseases that affect it
Ipmcenters.org/cropprofiles/docs/cawalnuts.pdf	General production information on this crop, the cultural practices involved, insect pests, diseases etc
Agroforestry.net/images/pdfs/Areca-catechu-betel-nut.pdf	Information on growing this crop which is cultivated in East Africa, South Asia and the Pacific
Https://rirdc.infoservices.com.au/downloads	Report from the Rural Industries Research and Development

/00-015.pdf	Corporation
Songonline.ca/establishing_orchard.htm	Information from the Society of Ontario Nut Growers on all aspects of establishing and caring

Olive
Top Websites in Olive Category

Website URL	Description
Internationaloliveoil.org	Contains the group's aims, activities, international agreements etc
Oliveoilpakistan.com	Project to study the suitability of the land and the environmental requirements for growing olives
Crfg.org/pubs/ff/olive.html	Brief botany, and notes on the cultivation of the tree, and the varieties available
Web.tiscali.it/OlivOlio/	Technical and professional consultancy for olive growing and olive oil, Tuscany, Italy
Iberianature.com/material/olives.html	Provides information on the history of this crop, its cultivation and the production

Pear
Top Websites in Pear Category

Website URL	Description
En.wikipedia.org/wiki/European_Pear	Article from Wikipedia on this fruit with information on its origins, cultivation etc
Resources.cas.psu.edu/TFPG/AGRS45part01-06.pdf	Information on site selection for this crop, soil preparation, varieties of Asian pears
Caf.wvu.edu/kearneysville/disease_descriptions/omfabrea.html	Information on this condition which can affect foliage and fruit, the symptoms, disease cycle etc
Ipmcenters.org/CropProfiles/docs/NewEnglandpear.pdf	General production information on this crop, the cultural practices involved, insect pests, diseases etc
Gardenaction.co.uk/fruit_veg_diary/pear_tree_1.asp:	Instructions on choosing and growing these fruit trees in the garden, with information on roots

Source: alexa.com

Prune Fruits
Top Websites in Prune Fruits Category

Website URL	Description
Beachplum.cornell.edu	Cornell University project to aid in establishing the fruiting shrub beach plum as a commercial variety
Resources.cas.psu.edu/TFPG/AGRS45part01-08.pdf	Information on site selection for this crop, soil preparation, growth regulators, cultivars
Resources.cas.psu.edu/TFPG/AGRS45part01-07.pdf	Information on site selection, soil preparation, nursery tree quality, cultivars, general care
Resources.cas.psu.edu/TFPG/AGRS45part02-07.pdf	Provides comprehensive information on this disease of stone fruit trees
Aggie-horticulture.tamu.edu/stonefruit/	Includes cultivar evaluations, articles, and news
Agroatlas.ru/cultural/Prunus_armeniaca_K_en.htm	Information on the apricot tree, its biology and morphology, distribution in the former USSR
Agroatlas.ru/cultural/Prunus_avium_K_en.htm	Information on the cherry tree, its biology and morphology, distribution in the former USSR
Ipmcenters.org/cropprofiles/docs/caplums.pdf	General production information on this crop, the cultural practices involved, insect pests, diseases etc
Gardenaction.co.uk/fruit_veg_diary/fruit_veg_mini_project_november_2_peach.asp	Information on growing and training this crop in cooler areas
Assuredproduce.co.uk/resources/000/255/703/Fruit_(Stone_Fruit)1.pdf	Producing crops in accordance with these standards allows growers to sell produce to any supermarket

Strawberry

Top Websites in Strawberry Category

Website URL	Description
Strawberryplants.org	The one stop for everything related to strawberry plants and growing strawberries
Strawberryplants.org/2010/05/growing-strawberries/	Comprehensive guide for the home gardener who wants to grow strawberries
Indiaagronet.com/indiaagronet/crop%20info/strawberry.htm	India's leading website dealing with strawberry plants
Ncstrawberry.com	Provides information aimed at commercial strawberry growers
Strawberry.ifas.ufl.edu	University of Florida's Strawberry Lab
Fruit.cornell.edu/tfabp/strawanthracnose.pdf	Information on this disease caused by Colletotrichum acutatum or C. gloeosporioides
Nysipm.cornell.edu/organic_guide/strawberry.pdf	Comprehensive guidance for strawberry growers or farmers wishing to switch to organic strawberry
Ipm.ucdavis.edu/PMG/selectnewpest.strawberry.html	University of California agricultural management guidelines for control of strawberry pests
Ohioline.osu.edu/hyg-fact/3000/pdf/HYG_3209_08.pdf	Photographs of diseased fruit, the symptoms, disease development and management
Ohioline.osu.edu/hyg-fact/3000/pdf/HYG_3201_08.pdf	This condition may occur after excessive rainfall
Ohioline.osu.edu/hyg-fact/3000/pdf/HYG_3211_08.pdf	Photographs of diseased leaf and fruit, the symptoms, the causal organism and its life cycle
Ohioline.osu.edu/hyg-fact/1000/1424.html	Cultivation guide for this crop, with advice on cultivars, pests and diseases and general management
Ohioline.osu.edu/hyg-fact/3000/pdf/HYG_	Fact sheet on these diseases which do not generally cause significant

3015_08.pdf	economic damage
Ohioline.osu.edu/hyg-fact/3000/pdf/HYG_3012_08.pdf	Photograph of diseased plant, with illustrations of some other common root problems
Edis.ifas.ufl.edu/in713	Information on the development of resistance to insecticides and miticides and how to deal with them
Edis.ifas.ufl.edu/CV003	Information on proper use of fertilizer which is important for maximizing yield and fruit quality
Edis.ifas.ufl.edu/NG031	Information on the sting nematode
Strawberry.ifas.ufl.edu/fumigation%20vs%20nonfumigation.htm	Fumigation is used by growers to prevent soil-borne pest and disease problems in strawberries
Ufdc.ufl.edu/IR00002733/00001	Photographs and information on the mite, Neoseiulus californicus, its life cycle, biology etc
Edis.ifas.ufl.edu/topic_hs_strawberry	Information on the achievements made by scientists at the Florida Agricultural Experiment Station
Eppserver.ag.utk.edu/Extension/SBost/pubs/strawberry-diseases-02.pdf	A guide to aid in the identification, prevention, and treatment of common strawberry diseases
Cals.uidaho.edu/edComm/pdf/CIS/CIS0931.pdf	A discussion of the basic questions and concerns related to beginning a commercial strawberry
Www1.agric.gov.ab.ca/$department/deptdocs.nsf/all/prm2561	Information on aphids, clipper weevils, leaf rollers, mites, slugs, spittle bugs, root weevils etc
Www1.agric.gov.ab.ca/$department/deptdocs.nsf/all/faq6860	Information on the timing of herbicide applications and the products that are suitable for these crops
Fera.defra.gov.uk/plants/publications/documents/QIC31.pdf	Description of distribution, symptoms, sources, development, disease status.

Vegetables General
Top Websites in Vegetables Category

Website URL	Description
Carrotmuseum.co.uk	A virtual museum whose mission is to educate, inform and amuse visitors through the collection of vegetables
Vfic.tamu.edu	Part of the Horticultural Department of Texas A&M University set up to develop vegetable
Avrdc.org	Tropical vegetable research at the Asian Vegetable Research and Development Center
Vric.ucdavis.edu	Information on the cultivation, production and harvesting of vegetables in California for farmers
Omafra.gov.on.ca/engl ish/crops/hort/vegeta ble.html	All aspects of vegetable production are covered including factsheets, newsletters, crop updates etc
Usask.ca/agriculture/pl antsci/vegetable/index. htm	Provide commercial and hobby producers of Saskatchewan with up to date, locally relevant, produce info
Ipm.ucdavis.edu/PMG /crops-agriculture.html	Wide ranging approach on all aspects of integrated pest management (IPM) principles
Edis.ifas.ufl.edu/NG03 2	Information on nematodes which attack these crops, their biology and life history, symptoms etc
Hort.purdue.edu/newc rop/afcm/mungbean.h tml	Information on this crop, its history, uses, growth habits, environmental requirements etc
Vegedge.umn.edu	Provides advice on IPM
Bioengr.ag.utk.edu/Ext ension/ExtProg/Veget able/	Experimental Station which has launched a Vegetable Initiative
Cuestaroble.com	Provides international consulting services for greenhouse vegetable projects
Msucares.com/insects/	Some recommendations for insect

vegetable/index.html	control for commercial vegetables in Mississippi plus photos
Hort.purdue.edu/newc rop/afcm/jerusart.html	Factsheet on this crop, its history, uses, varieties and cultivation
Hort.purdue.edu/newc rop/afcm/adzuki.html	Information on this crop
Pubs.ext.vt.edu/catego ry/plant-diseases.html	Provides information on vegetable diseases
Edis.ifas.ufl.edu/WG0 29	Information on the mechanical and chemical means of managing weeds in muskmelon, cucumber, squash etc
Insect.pnwhandbooks. org	Detailed information on a wide range of vegetable pests
Hort.purdue.edu/newc rop/afcm/chickpea.ht ml	Factsheet on this crop, its history, uses, varieties and cultivation
Hort.purdue.edu/newc rop/id- 56/vinecrops.html	Production guide for growers with much information on diseases problems, susceptibilities etc
Hort.purdue.edu/newc rop/afcm/fababean.ht ml	Factsheet on this crop, its history, uses, varieties and cultivation
Hort.purdue.edu/newc rop/afcm/fieldbean.ht ml	Factsheet on this crop, its history, uses, varieties and cultivation
Hort.purdue.edu/newc rop/proceedings1993/ V2- 045.html#AZUKI%20 BEAN	Information on these members of the Fabaceae family that might have potential to be grown commercially
Hort.cornell.edu/exten sion/commercial/veget ables/index.html	Homepage for Cornell University
Nysaes.cornell.edu/rec ommends/18frameset. html	Recommended varieties, planting methods, fertility, harvesting and information on pests etc

Source: alexa.com

Asparagus

Website URL	Description
Ohioline.osu.edu/b826/index.html	A guide to producing asparagus
Michiganasparagus.org	Promotes the production and sale of the vegetable, and assists in research

Beans

Website URL	Description
Hort.purdue.edu/newcrop/afcm/mungbean.html	Information on this crop, its history, uses, growth habits, environmental requirements etc
Hort.purdue.edu/newcrop/afcm/adzuki.html	Information on this crop, its history, uses, growth habits, environmental requirements etc
Hort.purdue.edu/newcrop/afcm/chickpea.html	Factsheet on this crop, its history, uses, varieties and cultivation
Hort.purdue.edu/newcrop/afcm/fababean.html	Factsheet on this crop, its history, uses, varieties and cultivation
Hort.purdue.edu/newcrop/afcm/fieldbean.html	Factsheet on this crop, its history, uses, varieties and cultivation
Hort.purdue.edu/newcrop/proceedings1993/V2-045.html#AZUKI%20BEAN	Information on these members of the Fabaceae family
Agroatlas.ru/cultural/Phaseolus_vulgaris_K_en.htm	Information on this crop, its biology and morphology, ecology, distribution in the former USSR
Jeffersoninstitute.org/pubs/mung_beans_guide.pdf	An overview of this legume, also known as green gram or golden gram, its description, cultivation etc

Source: alexa.com

Brassicas

Website URL	Description
Edis.ifas.ufl.edu/NG024	Information on general IPM considerations, symptoms, damage, field diagnosis and sampling etc
Dpi.nsw.gov.au/__data/assets/pdf_file/0005/80168/Cabbage-growing---Primefact-90-final.pdf	Factsheet from the NSW Department of Primary Industries covering soil types, varieties etc
Omafra.gov.on.ca/english/crops/facts/85-043.htm	Details on brassica diseases from the Ontario Ministry of Agriculture
Www1.agric.gov.ab.ca/$department/deptdocs.nsf/all/agdex3509	The ministry enables the growth of a competitive, sustainable agriculture industry

Source: alexa.com

Cucurbits

Website URL	Description
Edis.ifas.ufl.edu/WG0 29	Information on the mechanical and chemical means of managing weeds in muskmelon, cucumber, squash etc
Hort.purdue.edu/newc rop/id-56/vinecrops.html	Production guide for growers with much information on diseases problems, susceptibilities etc
Nysaes.cornell.edu/rec ommends/18frameset. html	Recommended varieties, planting methods, fertility, harvesting and information on pests
Ipm.ucdavis.edu/PMG /selectnewpest.cucurbi ts.html	Provides information on possible biological and chemical controls for the diseases, insects, etc
Edis.ifas.ufl.edu/NG02 5	Information on general IPM considerations, symptoms, damage, field diagnosis and sampling etc
Ces.ncsu.edu/depts/pp /notes/Vegetable/vdin 011/vdin011.htm	Photographs and information on the causal agent, symptoms and management of this condition
Extension.uga.edu/pub lications/detail.cfm	Information for gardeners including pest management, vegetable growing, a newsletter etc
Agrsci.unibo.it/wchr/ wc1/basile.html	Paper given at the World Congress on Horticultural Research concerning the economics of growing this crop
Apsnet.org/publication s/commonnames/Page s/Curcubits.aspx	Provides the common names of the bacterial, fungal, nematode, viral and other diseases etc
Attra.ncat.org/attra-pub/downymildew.ht ml	This publication discusses cultural controls and alternative pesticides
Forestryimages.org/br owse/detail.cfm	Quality photographs of forest insects and disease organisms
Agroatlas.ru/cultural/ Cucumis_sativus_K_en	Information on this crop, its biology and morphology, ecology, distribution

.htm	in the former USSR
Ipmcenters.org/CropProfiles/docs/NJsquash.pdf	Production information on squashes grown in New Jersey
Assuredproduce.co.uk/resources/000/235/978/Courgettes_Marrows_Squash_and_Pumpkins_2007.pdf	Producing crops in accordance with these standards allows growers to sell produce to any supermarket
Assuredproduce.co.uk/resources/000/235/596/Cucumbers_2007.pdf	Adopting this protocol will enable producers to sell cucumbers to any supermarket in the UK
Lpl.arizona.edu/~bcohen/cucumbers/history.html	Brief account of the cultivated plant and its introduction to the Americas

Source: alexa.com

Peppers

Website URL	Description
Edis.ifas.ufl.edu/NG032	Information on nematodes which attack these crops, their biology and life history, symptoms etc
Edis.ifas.ufl.edu/PP104	Photographs and information on the causal agent, symptoms, disease cycle, epidemiology etc
Edis.ifas.ufl.edu/HS368	Information on the materials and methods used and the results of trials held at the North Florida
Edis.ifas.ufl.edu/topic_pepper_fertilization_and_nutrition	Information on research at the University of Florida on fertilizer usage and application
Extension.uga.edu/publications/detail.cfm	Information for gardeners including pest management, vegetable growing, a newsletter etc
Www1.agric.gov.ab.ca/$department/deptdocs.nsf/all/opp4523	Information on the cultivation of this crop under glass with illustrations of the steps involved

Organic Vegetables

Website URL	Description
Nysaes.cornell.edu/recommends/11frameset.html	This is part of Cornell's 'Integrated Crop and Pest Management Guidelines for Commercial Vegetables
Vric.ucdavis.edu/veg_info_topic/organic_production.htm	This page gives a list of links to PDF documents on organic growing
Eap.mcgill.ca/publications/eap3.htm	Scholarly article discussing the issues surrounding the use of synthetic pesticides and alternatives

Tomato

Website URL	Description
Edis.ifas.ufl.edu/NG032	Information on nematodes which attack these crops, their biology and life history, symptoms etc
Sgn.cornell.edu/about/tomato_sequencing.pl	Details of the project, together with the gene sequences, listed by cultivar and position
Vegetablemdonline.ppath.cornell.edu/factsheets/Tomato_List.htm	Information on a number of viral, fungal and bacterial diseases that may infect tomato crops
Ipm.ucdavis.edu/PMG/selectnewpest.tomatoes.html	Provides an integrated pest management program from planting through to harvesting
Ohioline.osu.edu/hyg-fact/3000/3114.html	Factsheet with photographs and information on this condition, including the symptoms etc
Ohioline.osu.edu/hyg-fact/3000/3120.html	Photographs of these diseases affecting this crop, the symptoms, causal organisms and management
Ohioline.osu.edu/hyg-fact/3000/3112.html	Photographs of this disease, the symptoms, causal organism and control methods
Apsnet.org/publications/commonnames/Pages/Tomato.aspx	Provides the common names of the bacterial, fungal, nematode, viral and other diseases
Attra.ncat.org/attra-pub/tomato.html	Describes the methods, economics and issues in commercial organic production
Forestryimages.org/browse/detail.cfm	Quality photographs of forest insects and disease organisms
Forestryimages.org/browse/subimages.cfm	Quality photographs of forest insects and disease organisms
Agroatlas.ru/cultural/Lycopersicon_lycopersicum_K_en.htm	Information on this crop, its biology and morphology, ecology, distribution in the

	former USSR
Appliedbio-nomics.com/sites/default/files/410-tomato.pdf	This article outlines the biological agents available for controlling whitefly, fungus gnats, etc
Assuredproduce.co.uk/resources/000/255/526/Tomatoes_(protected)1.pdf	Producing crops in accordance with these standards allows growers to sell produce to any supermarket
Kdcomm.net/~tomato/	Dedicated to the art of growing tomatoes and includes details on seed sources, growing tips

Source: alexa.com

Vegetables-Pests and Diseases

Website URL	Description
Ipm.ucdavis.edu/PMG/crops-agriculture.html	Wide ranging approach on all aspects of integrated pest management (IPM) principles
Vegedge.umn.edu	Provides advice on all aspects of vegetable crop integrated pest management (IPM)
Msucares.com/insects/vegetable/index.html	Some recommendations for insect control for commercial vegetables in Mississippi plus photos
Pubs.ext.vt.edu/category/plant-diseases.html	Provides information on diseases that attack tomatoes, cucurbits, peppers and brassicas
Insect.pnwhandbooks.org	Detailed information on a wide range of vegetable pests relevant to the Pacific North West
Vric.ucdavis.edu/veg_info_topic/diseases.htm	Notes on a number of vegetable diseases
Edis.ifas.ufl.edu/pdffiles/CV/CV11200.pdf	Information on the microscopic parasitic roundworms that live in soil and attack plant roots etc
Edis.ifas.ufl.edu/TOPIC_BOOK_Plant_Disease_Management_Guide	Guides on controlling diseases in vegetables listed by crop
Ces.ncsu.edu/depts/pp/notes/Vegetable/vegetable_contents.html	Details on a range of diseases on a variety of herbs and vegetables including cucurbits, tomato etc
Extento.hawaii.edu/kbase/crop/crops/vegetabl.htm	Information on the biology and control of pests and diseases of vegetables plus other crops
Omafra.gov.on.ca/english/crops/insects/diseases.html	Fact Sheets on a variety of crop diseases including vegetables
Eap.mcgill.ca/CPCM_9.htm	Information on the lifecycle of the cabbage root fly, the damage it causes
Ipm.uconn.edu/pa_vegetable/	Provides information for growers of tomatoes, peppers, beans, cole

	crops, corn, cucurbits, potatoes etc
Ca.uky.edu/entomology/dept/entfacts2.asp#veg	Information on a range of insect pests and also beneficial insects

Source: alexa.com

Greenhouse Production or Protected Cultivation

Website URL	Description
Vric.ucdavis.edu/veg_info_topic/greenhouse.htm	Page contains links to several PDF documents including "Year Round Gardening with a Greenhouse"
Edis.ifas.ufl.edu/TOPIC_BOOK_Florida_Greenhouse_Vegetable_Production_Handbook	Information on setting up and managing a vegetable production enterprise under glass
Mrec.ifas.ufl.edu/Foliage/entomol/ncstate/thripkey.htm	Key to the species of thrips found in greenhouses in Florida
Extension.uga.edu/publications/detail.cfm	Information for gardeners including pest management, vegetable growing etc
Pods.dasnr.okstate.edu/docushare/dsweb/Get/Document-1281/F-6711web.pdf	Provides guidance on identifying insect, mite and disease problems
Www1.agric.gov.ab.ca/$department/deptdocs.nsf/all/agdex1443	The ministry enables the growth of a competitive, sustainable agriculture industry
Ghvi.co.nz	Articles and forums on crop management, disease control, nutrient film technique, and media etc

Source: alexa.com

TOP WEBSITES IN ORGANIC AGRICULTURE CATEGORY

Organic Farming and Sustainable Agriculture
Top Websites in Organic Farming/Sustainable Agriculture Category

Website URL	Description
Cgiar.org	With a mission to contribute to food security and poverty eradication in developing countries
Cornucopia.org	Research and news for consumers, farmers, and the media about primarily US-based organic and sustainable agriculture
Groworganic.com	Supplies tools, equipment, and goods to organic gardeners and farmers
Ebfarm.com	Organic food company
Soilassociation.org	Food you can trust, annotated links and information explaining how clean soil makes healthy foods
Fibl.org	Independent, non-profit, research institute for advancing cutting-edge science in the field of organic agriculture
Sare.org	Information on sustainable agriculture for producers
Sustainableagriculture.net	Dedicated to educating the public on the importance of a sustainable food and agriculture systems
Orgprints.org	Open access archive for papers related to research in organic agriculture
Omri.org	List materials approved for organic use in the USA
Attra.org	Provides technical assistance, news, funding resources, and

	sustainable agriculture publication
Rodaleinstitute.org	Website on world-renown organic research farm
Ifoam.org	An umbrella organisation for the national organic certification bodies
Ota.com	Represents the organic industry in Canada and the United States
Rspo.org	International organization of producers, distributors, conservationists and other stakeholders
Organic-bio.com	Database containing more than 2000 product groups and more than 14,000 addresses of producers
Acresusa.com	Monthly magazine covering commercial-scale organic and sustainable agriculture
Biodynamics.com	U.S. non-profit organization formed to foster knowledge of biodynamic methods of agriculture
Ccof.org	Since 1973, CCOF has been a leader in the organic industry
Sustainweb.org	Advocates food and agriculture policies and practices that enhance the health and welfare of people
Seattletilth.org	Promotes organic gardening in an urban setting
Isealalliance.org	Describes objectives, which are to improve social auditing processes in agriculture
Afsic.nal.usda.gov	USDA, National Agricultural Library site about sustainable and organic food production

	systems
Bountifulgardens.org	Offering seed for untreated heirloom and open pollinated vegetables, flowers, etc
Cias.wisc.edu	A sustainable agriculture research center at the University of Wisconsin

Source: alexa.com

Vermicomposting

Website URL	Description
Happydranch.com	Provides information on worms and vermicomposting
Howtocompost.org	Offering composting information and links to web resources on a variety of composting topics
Whatcom.wsu.edu/ag/compost/Redwormsedit.htm	An informative "how to" of home vermiculture
Working-worms.com	Provides information on the benefits of worm farming and an illustrated step by step guide
Bucketofworms.co.uk	Provides information on the Dendrobaena worm, composting, wormeries etc
Dirtmaker.com	Manufactures and sells flow-through worm bins for composting with worms
Wormwigwam.com	Procedures for composting using earthworms, and a vermicomposting system offered for sale
Calrecycle.ca.gov/Organics/Worms/	Resources for people seeking information about worm composting/vermicomposting, worm suppliers etc
Alabamajumpers.com	Offers worms for sale and a wholesale drop ship program
Wormfarmingsecrets.com	Free newsletter produced by experienced worm farmers and professionals in the vermiculture industry
Wormdigest.org	Quarterly newspaper that reports on worms and worm composting (vermicomposting) on all levels
Cathyscomposters.com	Canadian supplier of red wiggler worms, vermicomposting bins, and informative books
Vermicomposting.com	Blog about worms, castings and composting
Worms.blat.co.za	Blog on worm farming from a worm supplier in South Africa

Goldenrod.net/welcome.htm	Free six-week online course in worm composting
Bigsteamypile.com	Provides information on composting, vermicomposting and composting toilets etc
Groups.yahoo.com/group/the_worm_bin/	An active Yahoo group dedicated to vermicomposting
Thewormexpert.com	Open discussion forum for red worm composting. Ask questions and assist others
Wormtec.com.au	Provides information on worm farming and vermiculture, and works with schools and groups
Morarkango.com/biotechnology/research.php	Indian NGO provides information on vermiculture, vermicomposting and organic farming

Source: alexa.com

Waste Management

Website URL	Description
Biontech.com	Offers farmers a patented waste management process that reduces odors associated with animal waste management
Agronext.iastate.edu/immag/	IMMAG provides information on the coordinated effort to improve manure management
Epa.gov/waterscience/guide/cafo/	Text and background of U.S. Environmental Protection Agency regulations covering wastewater disposal
Wastemgmt.ag.utk.edu	Information on extension and research programs and publications, and on certification programs

Source: alexa.com

TOP WEBSITES IN CEREALS CATEGORY

Corn/Maize

Website URL	Description
En.wikipedia.org/wiki/Maize	Information from Wikipedia on this crop, its physiology, genetics, origin, cultivation, pests etc
Corn.org	U.S. national association representing the corn refining (wet milling) industry
Ncga.com	Represents America's corn growers
Agry.purdue.edu/ext/corn/	Corn (maize) information for corn growers throughout North America
Iowacorn.org	Provides research, science and market information
Commodityclassic.com	The combined educational event for the America Soybean Association and the National Corn Growers
Mncorn.org	Promoting the profitability of corn producers through support of expanded uses of corn
Ilcorn.org	A joint project between the Illinois Corn Growers Association and the Illinois Corn Marketing Board
Michag.com	Michigan Agricultural Commodities trades and handles corn, soybeans, wheat, grain, etc
Agron.missouri.edu	Developed at the University of Missouri to provide access to Maize-related genetic information
Necga.org	Highlights research, promotion and public policy efforts for growing corn in the state
Cepm.org	Description of missions and objectives, list of its members, and production key figures
Kycorn.org	Provides consumers and producers the latest news and information
Coloradocorn.com	Information regarding corn growers in Colorado
Sdcorn.org	Dedicated to serving farmers,

	educators, business people and others across the state
Ipm.ncsu.edu/corn/diseases/corn_diseases.html	A review of pest and diseases affecting this state's crops
Mocorn.org	Provides information on growing corn in Missouri
Texascorn.org	Includes news, calendar of events, research, and corn use information
Hort.purdue.edu/newcrop/afcm/popcorn.html	Factsheet on this crop, its history, uses, varieties and cultivation
Profiles.nlm.nih.gov/LL/Views/Exhibit/narrative/biographical.html	Information on the scientist and the work she carried out investigating the origins of maize
Silagebreeding.agronomy.wisc.edu	Information on corn silage analysis, breeding, and evaluation of germplasm available to the public
Ohioline.osu.edu/hyg-fact/3000/3119.html	Photographs of this common disease, the symptoms, causal organism and management
Edis.ifas.ufl.edu/IN302	Photographs and information on the life cycle of this pest, its description, host plants, etc
Edis.ifas.ufl.edu/IG060	Describes the insect pests most commonly encountered in this crop
Corn2.agron.iastate.edu	Detail of individual research programs at Iowa State University

Source: alexa.com

Oats

Website URL	Description
Wheat.pw.usda.gov/gg pages/oatnewsletter/	Sponsored by the American Oat Workers Conference
Ent.iastate.edu/imageg al/plantpath/oats/	Photographs of oats infected by crown rust and yellow dwarf virus, with links to articles
Www1.agric.gov.ab.ca/ $Department/deptdocs .nsf/All/prm7780	The ministry enables the growth of a competitive, sustainable agriculture industry
Apsnet.org/publication s/commonnames/Page s/Oats.aspx	Provides a list of the common and scientific names for the bacterial, fungal, viral and parasitic infestations
Forestryimages.org/br owse/detail.cfm	Quality photographs of forest insects and disease organisms

Barley

Website URL	Description
Barleyworld.org	Provides information on the Oregon State University breeding programs
Albertabarley.com	Enables Alberta, Canada farmers to invest in positive initiatives that promote the barley industry
Wheat.pw.usda.gov/ggpages/b gn/	Informal reports to further the exchange of ideas and information between research workers
Wheat.pw.usda.gov/ggpages/B arleyNewsletter/	A paper and online newsletter for informal communication between researchers on barley
Ent.iastate.edu/imagegal/plant path/barley/1785.15ergotinbarl ey.html	Photograph of infected seed and a link to an article on integrated crop management
Fao.org/fileadmin/user_upload /inpho/docs/Post_Harvest_C ompendium_-_BARLEY.pdf	Provides information on pre-harvest operations, harvesting, transport, threshing, drying, cleaning etc

Www1.agric.gov.ab.ca/$Department/deptdocs.nsf/all/prm2477	Information on Limothrips denticornis, its life cycle, economic importance and possible management
Www1.agric.gov.ab.ca/$department/deptdocs.nsf/all/prm7772	Several photographs of affected plants and suggested management strategies
Www1.agric.gov.ab.ca/$Department/deptdocs.nsf/All/prm7770	The ministry enables the growth of a competitive, sustainable agriculture industry
Apsnet.org/publications/commonnames/Pages/Barley.aspx	Provides the common names of the bacterial, fungal, nematode, viral and other diseases
Gbif.org/dataset/fa609d5b-f9e9-487b-803e-b3ff77e15790	Scientists from INRA (French national agronomic research institute)
Forestryimages.org/browse/detail.cfm	Quality photographs of forest insects and disease organisms
Agricrops.org	A database of British Barley Varieties

Source: alexa.com

Millets

Website URL	Description
Worldbank.org/html/cgiar/newsletter/Mar96/4millet.htm	Overview of the increasing production of millet in the continent
Hort.purdue.edu/newcrop/afcm/millet.html	Factsheet on this crop, its history, uses, varieties and cultivation
Vegparadise.com/highestperch29.html	Notes on the history from early times of growing millet, together with some recipes

Source: alexa.com

Pearl Millet

Website URL	Description
En.wikipedia.org/wiki/List_of_pearl_millet_diseases	Common and scientific names of the bacterial, fungal and viral diseases
Ars.usda.gov/Research/docs.htm	The main in-house research arm of the U.S. Department of Agriculture
Caes.uga.edu/Publications/pubDetail.cfm	Agricultural and applied economics department web site with information for prospective students
Apsnet.org/publications/commonnames/Pages/PearlMillet.aspx	Provides the common names of the bacterial, fungal, nematode, viral and other diseases
Jeffersoninstitute.org/pubs/pearl_millet_guide.pdf	An overview of this crop, its description, cultivation, harvesting, storage, utilization, marketing etc

Source: alexa.com

Rice
Top Sites in Rice Category

Website URL	Description
Irri.org	Non-profit agricultural research and training center
Pinoyrkb.com	Dynamic crop management system designed for Philippine irrigated lowland rice
Ricehoppers.net	Aims to provide a platform for knowledge sharing on issues and develop sustainable ways
Shigen.nig.ac.jp/rice/oryzabase/	Integrated rice genetic map and mutant database at the National Institute of Genetics in Japan
Warda.cgiar.org	An autonomous intergovernmental research association working to increase the sustainable production
Asiarice.org	Working to secure a prosperous future for Asia's rice societies by supporting rice education
Oisat.org/pests/diseases/fungal/rice_blast.html	Photograph of affected foliage, notes on the symptoms, the conditions that favor its development
Hort.purdue.edu/newcrop/afcm/wildrice.html	Factsheet on this crop, its history, uses, varieties and cultivation
Ipm.ucdavis.edu/PMG/selectnewpest.rice.html	Provides general information on pesticides and details of the diseases, invertebrates and weeds
Fao.org/ag/AGP/AGPC/doc/riceinfo/Riceinfo.htm	This organisation aims to provide information to assist farmers and institutions in their efforts
Agebb.missouri.edu/rice/diseases.htm	Description, control and management of diseases

Alamy.com/search-results.asp	Royalty Free and Rights Managed stock photography
Commondreams.org/headlines/100900-01.htm	An article on the status of US patents on Basmati rice, by Ranjit Devraj
Rgp.dna.affrc.go.jp	Part of the Japanese MAFF Genome Research Project. It is a joint plant genome research program
Cgiar.org/impact/research/rice.html	Information on the crop, its global significance, and current research
Apsnet.org/publications/commonnames/Pages/Rice.aspx	Provides the common names of the bacterial, fungal, nematode, viral and other diseases
Web.idrc.ca/en/ev-9299-201-1-DO_TOPIC.html	An account of the methods by which freshwater fish farming can be integrated with paddy rice growth
Forestryimages.org/browse/detail.cfm	Quality photographs of forest insects and disease organisms
Naturland.de/fileadmin/MDB/documents/Publication/English/rice.pdf	Covers all aspects of plant cultivation, biological methods of plant protection, harvesting etc
Carleton.ca/~bgordon/Rice/papers/zhimin99.htm	A detailed account supporting a hypothesis about the original domestication of rice

Source: alexa.com

Wheat
Top Sites in Wheat Category

Website URL	Description
En.wikipedia.org/ wiki/Wheat	Introductory article, including material on history, breeding and genetics, economics and production etc
Einkorn.com	Einkorn (Triticum monococcum) is considered to be the oldest form of cultivated wheat
Uswheat.org	Supports the sale of wheat by offering education for overseas buyers, onsite training services
Wheatworld.org	Wheat facts, marketing, and research information
Usask.ca/agricultu re/plantsci/winter _cereals/	A detailed online manual for growing winter wheat, particularly in western Canada
Wheatgenome.org	The IWGSC was set up to sequence the wheat genome so as to enhance our knowledge of its structure
Smallgrains.org	Maintained cooperatively by the Minnesota Association of Wheat Growers (MAWG)
Ksre.ksu.edu/whe atpage/	Provide links to other useful information about wheat in Kansas, the USA and the world
Wheatgrowers.ca	Non-profit, voluntary farm organization representing the interests of its members to government
Hort.purdue.edu/ newcrop/proceedi ngs1999/v4-182.html	Details of this little grown relative of durum wheat
Hort.purdue.edu/ newcrop/afcm/sp elt.html	Factsheet on this crop, its history, uses, varieties and cultivation
Ohioline.osu.edu/ ac-fact/0010.html	Photographs of lesions on wheat leaves, and information on Blumeria graminis
Edis.ifas.ufl.edu/I	Describes the insect pests most commonly

G067	encountered in this crop, with information on scouting
Cropandsoil.oregonstate.edu/wheat/	Provides a synopsis of research at the University
Ent.iastate.edu/imagegal/plantpath/wheat/	Photographs of plants infected with yellow dwarf virus, leaf rust, molds, powdery mildew etc

Source: alexa.com

Rye
Top Sites in Rye Category

Website URL	Description
En.wikipedia.org/wiki/Ergot	Illustrated article from Wikipedia on Claviceps purpurea which is parasitic on rye
Hort.purdue.edu/newcrop/afcm /rye.html	Features history, uses, growth habits, environment requirements, cultural practices, and varieties
Ohioline.osu.edu/agf-fact/0026.html	Factsheet from Ohio State University
Mda.state.mn.us/en/protecting/ sustainable/mfo/mfo-fruit-veg/rye.aspx	Production history and links from the Minnesota Department of Agriculture
Www1.agric.gov.ab.ca/$departm ent/deptdocs.nsf/all/prm7794	Some notes on identifying ergot, cottony snow mold and stem smut, and some possible management
Www1.agric.gov.ab.ca/$departm ent/deptdocs.nsf/all/agdex1269	Production information from the Alberta, Canada agricultural department
Apsnet.org/publications/commo nnames/Pages/Rye.aspx:	Provides a list of the common and scientific names of the bacterial, fungal, viral and parasitic infestations
Agmrc.org/commodities__produ cts/grains__oilseeds/rye.cfm	Information about the cereal form the US Agriculutral Marketing Research Center
Agmrc.org/media/cms/bushuk_ C8B79BAB55BB0.pdf	Article on this important crop by W Bushuk
Attra.org/attra-pub/rye.html	Discusses how to grow and use rye as a cover crop

Source: alexa.com

TOP WEBSITES IN LEGUME OR PULSE CROP CATEGORY

Legumes/Pulses
Top Sites in Legumes Category

Website URL	Description
Peanut-institute.org	Features research, recipes, FAQs, and educational materials
Soyatech.com	Soybean and oilseed resource and information for the food and feed industries
Unitedsoybean.org	National soybean checkoff farmer-led organization
Indianasoybean.com	Conducts soybean promotion and research
Soygrowers.com	ASA develops legislative efforts to improve soybean farmer profitability
Saskpulse.com	Represents over 14,000 pulse crop farmers in the province
Texaspeanutboard.com	Features research, news, and nutrition
Nsrl.uiuc.edu	Research areas include enhancing soy productivity, creating food uses, commodity markets
Nebraskasoybeans.org	Lists programs, announcements, and research information
Hort.purdue.edu/newcrop/articles/ji-cowpea.html	Information on this crop with a plant description, uses, markets, economics and cultivation
En.wikipedia.org/wiki/List_of_lentil_diseases	Gives the common and scientific name for the fungal, parasitic nematode and viral diseases
Pp4sd.org.uk/downloads/pdf/Case%20study%20peas.pdf	This processor of peas in the UK describes its policy on sustainable agricultural development
Hort.purdue.edu/newcrop/cropfactsheets/Chickpea.html	Purdue University factsheet on the cultivation, usage and botany of the crop
Hort.purdue.edu/newcrop/afcm/cowpea.html	Factsheet on this crop, Vigna unguiculata, its history, uses,

	varieties and cultivation
Hort.purdue.edu/newcrop/ afcm/drypea.html	Factsheet on this crop, its history, uses, varieties and cultivation
Hort.purdue.edu/newcrop/ afcm/guar.html	Info on Cyamopsis tetragonoloba which is a drought-tolerant annual legume
Hort.purdue.edu/newcrop/ afcm/lentil.html	Detailed factsheet on the growing and use of the crop
Hort.purdue.edu/newcrop/ afcm/peanut.html	Factsheet on this crop, its history, uses, varieties and cultivation
Hort.purdue.edu/newcrop/ duke_energy/Vicia_faba.ht ml	Notes on botany, cultivation, food value, and commercial production
Soydiseases.illinois.edu	Publications, disease descriptions, and links regarding soybean diseases
Soydiseases.uiuc.edu/index. cfm	General information about the departments and contacts
Vegetablemdonline.ppath.co rnell.edu/factsheets/Beans_ Anthracnose.htm	Fact sheet on this major disease of beans caused by the fungus Colletotrichum lindemuthianum
Nysaes.cornell.edu/recomm ends/13frameset.html	Information on growing these crops
Soybeans.umn.edu	Information on soybean diseases, insects, varieties from the University of Minnesota
Ohioline.osu.edu/hyg-fact/3000/3110.html	Photographs of beans affected with these diseases, the symptoms, causal organisms and management

Peanut
Top Sites in Peanut Category

Website URL	Description
Peanut-institute.org	Features research, recipes, FAQs, and educational materials
Texaspeanutboard.com	Features research, news, and nutrition
Hort.purdue.edu/newcrop/afcm/peanut.html	Factsheet on this crop, its history, uses, varieties and cultivation
Edis.ifas.ufl.edu/PI044	Production facts on this crop, production practices, and advice on pests, diseases and weeds
Edis.ifas.ufl.edu/IN176	An illustrated guide to recognizing the large number of pests which may attack this crop
Edis.ifas.ufl.edu/IG062	Information on integrated pest management for this crop
Edis.ifas.ufl.edu/AA258	Instructions on growing this crop, fertilizer use, planting, tillage, pest management and monitoring
Edis.ifas.ufl.edu/NG016	Information on the species of nematode that attack peanuts and their diagnosis and management
Pods.dasnr.okstate.edu/docushare/dsweb/Get/Document-2323/F-7186web.pdf	Photographs and much information on southern blight and sclerotina blight, the disease cycle, etc
Cgiar.org/impact/research/groundnut.html	Description, statistics, global economic and dietary significance, and current research

Source: alexa.com

Soybean
Top Sites in Soybean Category

Website URL	Description
Soyatech.com	Soybean and oilseed resource and information for the food and feed industries
Unitedsoybean.org	National soybean checkoff farmer-led organization
Indianasoybean.com	Conducts soybean promotion and research
Soygrowers.com	ASA develops legislative efforts to improve soybean farmer profitability
Nsrl.uiuc.edu	Research areas include enhancing soy productivity, creating food uses, commodity markets, etc
Nebraskasoybeans.org	Lists programs, announcements, and research information
Soydiseases.illinois.edu	Publications, disease descriptions, and links regarding soybean diseases
Soydiseases.uiuc.edu/index.cfm	General information about the departments and contacts
Soybeans.umn.edu	Information on soybean diseases, insects, varieties from the University of Minnesota
Edis.ifas.ufl.edu/IG064	Describes the insect pests most commonly encountered in this crop
Edis.ifas.ufl.edu/NG018	Information on the species that attack soybean, their diagnosis and management
Ams.usda.gov/AMSv1.0 /ams.fetchTemplateData .do	Includes six commodity divisions - Cotton, Dairy, Fruit and Vegetable, Livestock and Seed etc
Ipm.iastate.edu/ipm/icm /taxonomy/term/260	Articles on soybean beetles and aphids
Ipm.iastate.edu/ipm/icm /taxonomy/term/578	Articles on these pests with information on chemical and biological control
Ent.iastate.edu/imagegal	Photographs of plants infected by

/plantpath/soybean/	various pathogens and pests, with links to articles
Extension.agron.iastate.edu/soybean/topicpage2.html	Photographs and information from Iowa State University
Attra.ncat.org/attra-pub/whitemold.html	This article discusses non-chemical options for organic control of white mold on soybeans
Forestryimages.org/browse/detail.cfm	Quality photographs of forest insects and disease organisms
Plantmanagementnetwork.org/edcenter/seminars/SoybeanIntro/	A short video to help plant practitioners improve the health, management, and production of soybeans
Arkansassoybean.com	Organization working on behalf of soybean producers in Arkansas

Source: alexa.com

Pea
Top Sites in Pea Category

Website URL	Description
Pp4sd.org.uk/downloads/pdf/Case%20study%20peas.pdf	This processor of peas in the UK describes its policy on sustainable agricultural development
Hort.purdue.edu/newcrop/afcm/cowpea.html	Factsheet on this crop, Vigna unguiculata
Hort.purdue.edu/newcrop/afcm/drypea.html	Factsheet on this crop
Vegetables.wsu.edu/peareport.html	Research project studying various varieties of peas to evaluate the use of the leaves and shoots
Ianrpubs.unl.edu/epublic/live/ec187/build/ec187.pdf	Information on this crop
Ext.nodak.edu/extpubs/plantsci/rowcrops/a1166w.htm	Notes on uses, adaptation, varieties, performance, field selection, seeding rates, inoculation etc
Unilever.com/Images/2003%20Sustainable%20Vining%20Peas%20-%20-%20Good%20Agricultural%20Practice%20Guidelines_tcm13-5323.pdf	Guidelines for farmers on good agricultural practice
Www1.agric.gov.ab.ca/$department/deptdocs.nsf/all/prm7819	An identification guide to the diseases that may infect this crop, with many photographs
Msucares.com/pubs/publications/p1535.htm	Information on cowpeas
Regional.org.au/au/asa/2001/2/a/peck1.htm	Includes a study of the practices, principal problems, and a discussion of its relative importance

Source: alexa.com

Bean – Phaseolus
Top Sites in Bean – Phaseolus Category

Website URL	Description
Vegetablemdonline.ppath.cornell.edu/factsheets/Beans_Anthracnose.htm	Fact sheet on this major disease of beans caused by the fungus Colletotrichum lindemuthianum
Nysaes.cornell.edu/recommends/13frameset.html	Information on growing these crops including recommended varieties, planting methods, fertility etc
Ohioline.osu.edu/hyg-fact/3000/3110.html	Photographs of beans affected with these diseases, the symptoms, causal organisms and management
Www1.agric.gov.ab.ca/$department/deptdocs.nsf/all/prm7818#anthra	An identification guide to various diseases that may attack this crop
Forestryimages.org/browse/detail.cfm	Quality photographs of forest insects and disease organisms

Source:alexa.com

Chick Pea
Top Sites in Chick Pea Category

Website URL	Description
Hort.purdue.edu/newcrop/cropfactsheets/Chickpea.html	Purdue University factsheet on the cultivation, usage and botany of the crop
Www1.agric.gov.ab.ca/$department/deptdocs.nsf/all/prm7838#botrytis	A guide to identifying botrytis blight and ascochyta blight, and possible management strategies

Source: alexa.com

Lentils
Top Sites in Lentils Category

Website URL	Description
En.wikipedia.org/wiki/List_o f_lentil_diseases	Gives the common and scientific name for the fungal, parasitic nematode and viral diseases
Hort.purdue.edu/newcrop/af cm/lentil.html	Detailed factsheet on the growing and use of the crop, from the Alternative Field Crops Manual
Www1.agric.gov.ab.ca/$depar tment/deptdocs.nsf/all/prm7 830#ab	An identification guide to the diseases that may infect this crop, with photographs
Cgiar.org/impact/research/le ntils.html	History of cultivation, statistics, and current research
Apsnet.org/publications/plan tdisease/backissues/Docume nts/1987Abstracts/PD_71_5 8.htm	Outlines research comparing different methods of preventing this disease

Source: alexa.com

TOP WEBSITES IN OIL SEEDS CATEGORY

Oilseeds General

Website URL	Description
Sunflowernsa.com	Information about the sunflower seed industry
Sunflowernsa.com/oil/	Promotional site for the oil product of sunflower
Hort.purdue.edu/newcrop /afcm/canola.html	Factsheet on this crop, its history, uses, varieties and cultivation
Hort.purdue.edu/newcrop /afcm/castor.html	Factsheet on this crop grown for its oil, its history, uses, varieties and cultivation
Hort.purdue.edu/newcrop /afcm/crambe.html	Factsheet on this crop, Crambe abyssinica, its history, uses, varieties and cultivation
Hort.purdue.edu/newcrop /afcm/flax.html	Factsheet on this crop, its history, its uses as an oilseed and as fibre, its varieties etc
Hort.purdue.edu/newcrop /afcm/meadowfoam.html	Factsheet on this crop, Limnanthes alba, its history, uses, varieties and cultivation
Hort.purdue.edu/newcrop /afcm/sesame.html	Factsheet on this crop, one of the oldest cultivated plants, its history, uses, varieties etc
Hort.purdue.edu/newcrop /afcm/sunflower.html	Factsheet on this crop, its history, uses, varieties and cultivation
Hort.purdue.edu/newcrop /afcm/vernonia.html	This potential oilseed crop, Vernonia galamensis, is native to eastern Africa
Fao.org/docrep/X5043E/ X5043E00.htm	Information from the FAO Agricultural Organization on the cultivation
Ianrpubs.unl.edu/epublic/ pages/publicationD.jsp	University of Nebraska-Lincoln publications on feeding, breeding, herd management etc
Www1.agric.gov.ab.ca/$D	Mamestra configurata is a

epartment/deptdocs.nsf/all/agdex3508	significant pest of canola
Www1.agric.gov.ab.ca/$Department/deptdocs.nsf/all/agdex2540	Photographs and information on the life history of this insect
Www1.agric.gov.ab.ca/$department/deptdocs.nsf/all/prm7710	The ministry enables the growth of a competitive, sustainable agriculture industry
Apsnet.org/publications/commonnames/Pages/Rapeseed.aspx	Provides the common names of the bacterial, fungal, nematode, viral and other diseases
Attra.org/attra-pub/oilseed.html	Describes the basic processes involved in oilseed processing
Ipmcenters.org/CropProfiles/docs/mncanola.pdf	Product information for this crop in Minnesota
Ipipotash.org/udocs/No%2016%20Oilseed%20rape.pdf	Introduction to this crop, the botany of the plant and its uses, and much information
Gmo-safety.eu/science/oilseed-rape/278.weeds-pests.html	Information on the production of this crop and the cultivation methods employed

Source: alexa.com

TOP WEBSITES IN FIBER CROPS CATEGORY

Fiber Crops General

Website URL	Description
Flaxcouncil.ca	National organization focused on promoting Canadian flax and flax products
Extension.oregon state.edu/catalog/ html/sb/sb681/	Covers botany, agronomy, soil characteristics, and other topics
Cottongrower.co m	Bi-monthly magazine for the US cotton growing industry with a library of articles
Archives.gov/edu cation/lessons/co tton-gin-patent/	Web site of the US National Archives providing classroom information for teachers
Sustainablecotton. org	USA. Non-profit organization, dedicated to the promotion of certified, organically grown cotton
Jute.org	Belgium. Intergovernmental organization under the aegis of UNCTAD
En.wikipedia.org/ wiki/Ramie	Information from Wikipedia on this plant, Boehmeria nivea, its cultivation, history, properties
En.wikipedia.org/ wiki/Kenaf	Information from Wikipedia on Hibiscus cannibinus and the fiber obtained from it
Cottonaustralia.co m.au	Information for members, the board, contact information and downloadable publications
Georgiacottonco mmission.org	Producer funded organization of cotton farmers in Georgia, USA
Cottondb.org	USA. Database containing genomic, genetic and taxonomic information for cotton
Cottonseed.com	Trade association. Includes information on membership, and cottonseed products
Evergreen-fs.com	Crop Scouting and Consulting services
Wcrl.ars.usda.gov	This site contains information on cotton research programs conducted or based at the USDA-ARS
Cottoninfo.ucdavi s.edu	Information resource for cotton growers in California
Ars.usda.gov/is/ AR/archive/aug0	Information on research into uses for the residues from this crop after the fiber has

0/kenaf0800.htm	been extracted
En.wikipedia.org/ wiki/List_of_cott on_diseases	Catalog of diseases that affect this crop from Wikipedia
Hort.purdue.edu/ newcrop/afcm/ke naf.html	History of kenaf, its uses, growth habits, environmental requirements, cultural practices etc
Hort.purdue.edu/ newcrop/nexus/ Hibiscus_cannabi nus_nex.html	Index of articles about kenaf and its cultivation
Hort.purdue.edu/ newcrop/duke_en ergy/Hibiscus_ca nnabinus.html	Section from the 1983 Handbook of Energy Crops covers the uses, folk medicine, chemistry etc
Hort.purdue.edu/ newcrop/proceedi ngs1993/v2- 402.html	Information on kenaf history, cultivations, resources
Hort.purdue.edu/ newcrop/proceedi ngs1996/v3- 060.html	Technical paper examining the current status of several potential industrial fiber crops
Utexas.edu/center s/nfic/	Information clearinghouse on the natural fibers, oilseeds, and related industries
Usda.mannlib.cor nell.edu/MannUs da/viewDocumen tInfo.do	Gives projections of expected market growth based on current trends of economic development
Ipm.ucdavis.edu/ PMG/selectnewp est.cotton.html	Provides an integrated pest management program from planting through to harvesting

Source: alexa.com

TOP WEBSITES IN FORAGE CROPS CATEGORY

Grassland, Hay and Forage

Website URL	Description
Forages.oregonstate.edu	Topics includes information on species, varieties, grazing, hay, and silage systems, management etc
Haytalk.com	A hay, forage and silage community with a forum, articles, blogs and news
Alfalfa.ucdavis.edu	Information on symposia proceedings, research, testing, seed production, and the alfalfa industry
Foragetesting.org	NFTA aims to improve the accuracy of forage testing and build grower confidence
Afgc.org	The AFGC sponsors an annual conference which is attended by forage and livestock producers
Pasture4horses.com	A technical and practical guide to the management and improvement of pasture
Britishgrassland.com	Information on the society and membership, news, special interest groups and local societies
Alfalfa.org	Develops publications and programs covering a wide range of subjects of concern to alfalfa growing
Nationalhay.org	Organization for those involved in the production, sale and transport of forage products
Ibiblio.org/farming-connection/grazing/home.htm	Information resources for efficient production of meat and milk from pastures
Forages.psu.edu	Links to web sites about forage

	production
Forages.psu.edu/topics/soil_fertility/	Pre-establishment, establishment and maintenance requirements of crops used for forage
Hort.purdue.edu/newcrop/afcm/grassseed.html	Factsheet on cool-season grasses and their cultivation and harvesting for seed
Hort.purdue.edu/newcrop/afcm/vetch.html	This legume is grown for soil improvement and for pasture
Hort.purdue.edu/newcrop/afcm/kochia.html	Kochia is grown as a drought-resistant forage crop for sheep and cattle
Hort.purdue.edu/newcrop/afcm/rutabaga.html	Factsheet on this crop, also known as swede, grown for human and animal consumption
Hort.purdue.edu/newcrop/afcm/forage.html	Forage sorghums are used primarily as silage for livestock
Hort.purdue.edu/newcrop/afcm/turnip.html	Factsheet on this crop, its history, its uses as a forage crop, its varieties and cultivation
Ipm.ucdavis.edu/PMG/r1300211.html	Both these aphids attack alfalfa and can cause damage
Alfalfa.ucdavis.edu/IrrigatedAlfalfa/pdfs/UCAlfalfa8302ForageQuality_free.pdf	A comprehensive report from the Alfalfa Workgroup at the University of California
Edis.ifas.ufl.edu/IG061	Describes the insect and mite pests most commonly encountered in forage crops
Ces.ncsu.edu/depts/pp/notes/oldnotes/ad1.htm	Provides information on disease identification, control, field monitoring, scouting and resistance etc
Ipm.iastate.edu/ipm/icm/2006/5-22/alfalfainsects.html	Photographs and information to help in identification of the many insects
Ipm.iastate.edu/ipm/icm/1998	Photographs and information

/5-4-1998/alfdis98.html	on a number of diseases that affect this crop
Ent.iastate.edu/imagegal/plant path/alfalfa/	Links to images and articles about diseases affecting alfalfa plants and crops

Source: alexa.com

TOP WEBSITES IN FIELD CROPS CATEGORY

Tobacco

Website URL	Description
Coffinails.com	Certified virus free Virginia and Havana tobacco seed for cigarettes and cigars
Edis.ifas.ufl.edu/IG066	Insects can cause considerable damage to this crop
Ces.ncsu.edu/depts/pp/notes/Tobacco/tobacco_contents.html	Provides information on the symptoms, factors affecting development, and control recommendation
Uky.edu/Agriculture/kpn/kyblue/kyblue.htm	An advisory system on tobacco blue mold
Apsnet.org/publications/commonnames/Pages/Tobacco.aspx	Provides the common names of the bacterial, fungal, nematode, viral and other diseases
Forestryimages.org/browse/detail.cfm	Quality photographs of forest insects and disease organisms
Tobaccofreekids.org/facts_issues/fact_sheets/toll/	Set of reports on tobacco farming, tobacco communities, and tobacco industry
Lookd.com/tobacco	Site with information on subjects including tobacco history, cultivation, aging, curing, etc
Tobaccoleaf.org	Trade organization open to all national growers' associations
Gaipm.org/top50/tobaccoflea.html	A description of this species with photographs and details of its hosts, damage done

Source: alexa.com

Sugarcane

Website URL	Description
Vsisugar.com	Information, Research and Development on sugarcane
Edis.ifas.ufl.edu/SC054	Comprehensive information on growing this crop including a variety fact sheet
Edis.ifas.ufl.edu/IG065	Information on a number of insect pests that may attack this crop
Edis.ifas.ufl.edu/IN529	Information on the species that attack sugarcane, their diagnosis and management
Edis.ifas.ufl.edu/IN210	Description of this pest which also attacks bananas and palms
Edis.ifas.ufl.edu/topic_book_ sugarcane_handbook	Information on the cultivation of sugarcane in Florida
Ars.usda.gov/main/site_main .htm	The main in-house research arm of the U.S. Department of Agriculture
Apsnet.org/publications/com monnames/Pages/Sugarcane. aspx	Provides the common names of the bacterial, fungal, nematode, viral and other diseases
Naturland.de/fileadmin/MD B/documents/Publication/E nglish/sugarcane.pdf	Covers all aspects of plant cultivation, biological methods of plant protection, harvesting
Caneinfo.nic.in	The Sugarcane Breeding Institute has developed this user-centered website
Isppweb.org/names_sugarcan e_common.asp	Information on the diseases that may affect sugarcane and their causal agents

Source: alexa.com

Sorghum

Website URL	Description
En.wikipedia.org/wiki/Sorghum	Account of the crop plant, and of its botany
Hort.purdue.edu/newcrop/afcm/broomcorn.html	Factsheet on this crop, a variety of sorghum, its history, uses, varieties and cultivation
Hort.purdue.edu/newcrop/afcm/sorghum.html	Factsheet on this crop, its history, uses, varieties and cultivation
Hort.purdue.edu/newcrop/afcm/syrup.html	Factsheet on this crop, its history, its uses as a substitute for sugar
Edis.ifas.ufl.edu/IG063	Describes the insect pests most commonly encountered in this crop
Ianrpubs.unl.edu/epublic/pages/publicationD.jsp	University of Nebraska-Lincoln publications on feeding, breeding, herd management etc
Cgiar.org/impact/research/sorghum.html	Describes programs of research into farming methods for this crop internationally
Apsnet.org/publications/commonnames/Pages/Sorghum.aspx	Provides the common names of the bacterial, fungal, nematode, viral and other diseases

Source: alexa.com

TOP WEBSITES IN ROOT CROPS
CATEGORY

Root Crops General

Website URL	Description
Idahopotato.com	Informational pages include recipes, photos, tips, and offers of genuine Idaho potatoes
En.wikipedia.org/wiki/Yam_(vegetable)	Describes the main cultivated species, preparation, use, and cultural aspects
Cipotato.org	Seeking to reduce poverty and achieve food security on a sustained basis in developing countries
Uspotatoes.com	Informational site about the National Potato Promotion Board and its programs
Potatoes.com	WSPC - Washington State Potatoes. Information on the commercial potato market and the vegetable
Potato.org.uk	A body that seeks to serve and develop the potato growing industry
Sbreb.org	Detailed information for growing sugar beet
Spudman.com	Online version of print magazine provides the latest industry news, an up to date calendar
Potatogrower.com	The official web site of Potato Grower Magazine
Nationalpotatocouncil.org	Non-profit trade association working for the well being of the U.S. potato industry
Isppweb.org/foodsecurity_cassavaghana.asp	Information on the causes and symptoms of these diseases etc
Mvproduce.com	Based in Colorado. Photographic tour of seed potato farm, potato variety growing guidelines etc
Europotato.org	Online listing of cultivated varieties of potato, their history and characteristics
Ces.ncsu.edu/depts/hort/hil/hil-23-a.html	The North Carolina State University Cooperative Extension Service explains all
Extento.hawaii.edu	Hawaii extension publication

/kbase/reports/swe etpot_prod.htm	
Potatobeetle.org	Information on biology and management of the Colorado potato beetle, Leptinotarsa decemlineata
Ces.ncsu.edu/depts /pp/notes/oldnote s/no103.htm	Information on this disease caused by the fungus Monilochaetes infuscans, and how to manage it
Hort.purdue.edu/n ewcrop/afcm/sugar beet.html	Factsheet on this crop, its history, uses, varieties and cultivation
En.wikipedia.org/w iki/List_of_cassava _diseases	Common and scientific names of the bacterial, fungal and viral diseases
Hort.purdue.edu/n ewcrop/duke_energ y/Ipomoea_batatas. html	Uses, description, and cultivation of the Sweetpotato
Vegetablemdonline. ppath.cornell.edu/ NewsArticles/Potat o_Virus.htm	Photographs of symptoms and information on virus Y, leafroll virus, virus S, virus X, etc
Sugarbeet.ucdavis.e du	Sugarbeet information, research, pest management, and cultivation
Ipm.ucdavis.edu/P MG/selectnewpest. sugarbeet.html	Provides general information on pesticides and details of the diseases, insects, mites, nematodes etc
Ipm.ucdavis.edu/P MG/selectnewpest. potatoes.html	Provides a year-round IPM program for potatoes with information
Ohioline.osu.edu/h yg- fact/3000/3103.ht ml	Photographs of this important disease of tubers, the symptoms, causal organism and management

Cassava

Website URL	Description
Isppweb.org/foodsecurity_cassavaghana.asp	Information on the causes and symptoms of these diseases, their means of spread etc
En.wikipedia.org/wiki/List_of_cassava_diseases	Common and scientific names of the bacterial, fungal and viral diseases
Edis.ifas.ufl.edu/MV042	Notes on this crop as grown in Florida, its uses and culture
Fadr.msu.ru/rodale/agsieve/txt/vol2/7/art2.html	A page on intecropping cassava with trees
Cgiar.org/impact/research/cassava.html	Information on the crop, its significance, and current research
Apsnet.org/publications/commonnames/Pages/Cassava.aspx	Provides the common names of the bacterial, fungal, nematode, viral and other diseases
New-ag.info/focus/focusItem.php	The New Agriculturist provides the latest news and information on agriculture and development
New-ag.info/01-1/develop/dev01.html	Cassava brown streak disease is a viral disease damaging the roots
Oisat.org/pests/diseases/viral/cassava_mosaic_disease.html	Photograph of affected foliage, and notes on the symptoms of this disease
Cassavabiz.org/production/pathology.htm	Photographs of affected plants and information on cassava brown streak virus disease

Source: alexa.com

Potato

Website URL	Description
Idahopotato.com	Informational pages include recipes, photos, tips, and offers of genuine Idaho potatoes
Cipotato.org	Seeking to reduce poverty and achieve food security on a sustained basis in developing countries
Uspotatoes.com	Informational site about the National Potato Promotion Board and its programs
Potatoes.com	WSPC - Washington State Potatoes. Information on the commercial potato market and the vegetable
Potato.org.uk	A body that seeks to serve and develop the potato growing industry
Spudman.com	Online version of print magazine provides the latest industry news, an up to date calendar
Potatogrower.com	The official web site of Potato Grower Magazine
Nationalpotatocouncil.org	Non-profit trade association working for the well being of the U.S. potato industry
Mvproduce.com	Based in Colorado. Photographic tour of seed potato farm, potato variety growing guidelines
Europotato.org	Online listing of cultivated varieties of potato, their history and characteristics
Potatobeetle.org	Information on biology and management of the Colorado potato beetle, Leptinotarsa decemlineata
Vegetablemdonline.ppath.cornell.edu/NewsArticles/Pot	Photographs of symptoms and information on virus Y, leafroll

ato_Virus.htm	virus, virus S, virus X, etc
Ipm.ucdavis.edu/PMG/sele ctnewpest.potatoes.html	Provides a year-round IPM program for potatoes with information
Ohioline.osu.edu/hyg-fact/3000/3103.html	Photographs of this important disease of tubers, the symptoms, causal organism and management
Ohioline.osu.edu/hyg-fact/3000/3106.html	Photographs of these diseases affecting potatoes, the symptoms, causal organisms and management
Ohioline.osu.edu/hyg-fact/3000/3107.html	Photographs of these diseases which cause serious losses in stored tubers, the symptoms etc
Ohioline.osu.edu/hyg-fact/3000/3104.html	Photographs of these diseases affecting tubers, the symptoms, causal organisms and management
Ohioline.osu.edu/hyg-fact/3000/3108.html	Photographs of these diseases affecting potatoes, the symptoms, causal organisms and management
Ohioline.osu.edu/hyg-fact/3000/3105.html	Photographs of potato tubers infected with this disease, the symptoms and causal organisms
Edis.ifas.ufl.edu/NG029	Information on general IPM considerations, symptoms, damage, field diagnosis and sampling etc
Ars.usda.gov/is/AR/archiv e/jan98/tool0198.htm	News on a friendly potato gene known as ubiquitin7
Fao.org/inpho/content/co mpend/text/ch17.htm#To pOfPage	Provides information on harvesting, transport, threshing, drying, cleaning, packaging and storage
Ianrpubs.unl.edu/epublic/li ve/ec1565/build/ec1565.pd f	Illustrated account of the various pests that attack this crop below the ground
Ndsu.nodak.edu/instruct/g udmesta/lateblight/	Disease management, photos of symptoms, information on other tuber and foliar diseases
Www1.agric.gov.ab.ca/$dep	The ministry enables the growth

| artment/deptdocs.nsf/all/a gdex11110/$file/625-4.pdf | of a competitive, sustainable agriculture industry |

Onion

Website URL	Description
Caes.uga.edu/Publications/p ubDetail.cfm	Agricultural and applied economics department web site with information for prospective student
Assuredproduce.co.uk/resour ces/000/255/645/Onions_(b ulb)1.pdf	Producing crops in accordance with these standards allows growers to sell produce to any supermarket

Sugar Beet

Website URL	Description
Sbreb.org	Detailed information for growing sugar beet
Hort.purdue.edu/newcrop/ afcm/sugarbeet.html	Factsheet on this crop, its history, uses, varieties and cultivation
Sugarbeet.ucdavis.edu	Sugarbeet information, research, pest management, and cultivation
Ipm.ucdavis.edu/PMG/sele ctnewpest.sugarbeet.html	Provides general information on pesticides and details of the diseases, insects, mites, nematodes etc
Edis.ifas.ufl.edu/IN262	Photographs and information on the life cycle of this pest, its description, host plants, damage etc
Nematode.unl.edu/extpubs /wyosbn.htm	Account of the damage caused by this pathogen, control measures, and photographs and drawings
Www1.agric.gov.ab.ca/$De partment/deptdocs.nsf/all/ agdex580	Description and life history of flea beetles, Psylliodes punctulata melsheimer, on sugar beet production
Apsnet.org/publications/co mmonnames/Pages/Beet.as px	Provides a list of the common and scientific names of the bacterial, fungal, viral and nematodes etc
Forestryimages.org/browse /detail.cfm	Quality photographs of forest insects and disease organisms
Ipmcenters.org/CropProfile s/docs/mtsugarbeet.pdf	Botanical description of the beet plant, its production statistics in Montana, cultural practices etc

Sweet Potato

Website URL	Description
Extento.hawaii.edu/kbase/reports/sweetpot_prod.htm	Hawaii extension publication
Ces.ncsu.edu/depts/pp/notes/oldnotes/no103.htm	Information on this disease caused by the fungus Monilochaetes infuscans, and how to manage it
Hort.purdue.edu/newcrop/duke_energy/Ipomoea_batatas.html	Uses, description, and cultivation of the Sweetpotato
Edis.ifas.ufl.edu/NG030	Information on general IPM considerations, symptoms, damage, field diagnosis and sampling etc
Entnemdept.ifas.ufl.edu/creatures/veg/potato/sweetpotato_weevil.htm	Information, management and references for this serious worldwide pest
Aggie-horticulture.tamu.edu/plantanswers/vegetables/sweetpotato.html	Facts and recipes
Apsnet.org/publications/commonnames/Pages/Sweetpotato.aspx	Provides the common names of the bacterial, fungal, nematode, viral and other diseases
Aces.edu/pubs/docs/A/ANR-0982/	Alabama extension publication
Forestryimages.org/browse/detail.cfm	Quality photographs of forest insects and disease organisms
Msucares.com/crops/comhort/sweetpotatoes.html	Mississippi extension publication, with information on cultivation techniques, management etc

Source: alexa.com

Yam

Website URL	Description
En.wikipedia.org/wiki/Yam_ (vegetable)	Describes the main cultivated species, preparation, use, and cultural aspects
Edis.ifas.ufl.edu/MV153	Notes on this crop, its uses, the species generally grown in Florida and their culture
Cgiar.org/impact/research/yam.html	Account of the importance and extent of cultivation of the crop today
New-ag.info/06-3/develop/dev03.html	Nigeria is the leading producer of yams in sub-Saharan Africa
Bspp.org.uk/icpp98/4.7/1.html	Outlines research into the nature and impact of diseases which may affect the yam crop

Source: alexa.com

TOP WEBSITES IN PLANTATION CROPS CATEGORY

Tea

Website URL	Description
Nhm.ac.uk/jdsml/nature-online/seeds-of-trade/page.dsml	Introduction to the history of cultivation and spread of some common crops
Food-info.net/uk/products/tea/cultivation.htm	Photographs and article describing how tea is grown
Naturland.de/fileadmin/MDB/documents/Publication/English/tea.pdf	Covers all aspects of plant cultivation, biological methods of plant protection, harvesting etc
Tea.co.uk/tea-growing-and-production	Provides information on growing, processing and blending tea
Ahmadtea.com/index.php/learn-about-tea/tea-manufacturing.html	Information on tea, how and where it is grown, processed and blended

Source: alexa.com

Cocoa

Website URL	Description
Worldcocoafoundation.org	Promotes a sustainable cocoa economy through economic and social development
Cocoainitiative.org	Partnership between NGOs, labour unions, cocoa processors and the major chocolate brands
Gutenberg.org/files/16035/16035-h/16035-h.htm	Online illustrated book by Brandon Head on cocoa, its growth and cultivation, its manufacture etc
Users.aber.ac.uk/gwg/pdf/griffith-wbdbiologist.pdf	Article by Gareth Griffith on these two pathogens that have caused havoc in the cacao industry
Icgd.rdg.ac.uk	Information on ICGD, based at the University of Reading, UK

Source: alexa.com

Coffee

Website URL	Description
Coffeeresearch.org/agriculture/soil.htm	Describes the nutrient requirements of coffee
Theatlantic.com/issues/99aug/9908ecocoffee.htm	Article from the Atlantic Monthly about sustainable coffee
Hort.purdue.edu/newcrop/duke_energy/Coffea_arabica.html#Cultivation	Information on all aspects of coffee production and use
Coffeeresearch.org/agriculture/diseases.htm	Provides information on coffee leaf rust, coffee berry disease, bacterial blight, nematodes
Naturland.de/fileadmin/MDB/documents/International/English/Coffee_Berry_Borer.PDF	Article discussing the main cultural and biological methods for controlling this pest
Ipmcenters.org/CropProfiles/docs/hicoffee.pdf	General information on growing coffee in Hawaii
Itdg.org/docs/technical_information_service/coffee.pdf	This article concentrates on the harvesting and processing of the crop
Thecoffeeproject.org.uk/coffee.php	Provides information on a project in Malawi to increase coffee production
Cafebar.co.uk/coffee_school/history_of_coffee/coffee_cultivation.aspx	Brief article on how coffee is grown
Indiaagronet.com/indiaagronet/crop%20info/coffee.htm	Information on cultivating this crop under the climatic conditions existing in India
Realcoffee.co.uk/Article.asp	Supplier of origin coffees and blends and speciality teas
Rombouts.com/uk/coffee/culture.html	Information on cultivating this crop
Coffeefacts.com	Provides information on the coffee bean, the plant, its cultivation, roasting, processing etc

TOP WEBSITES IN DAIRY SCIENCE AND ANIMAL HUSBANDRY CATEGORY

Poultry

Website URL	Description
Thepoultrysite.com	Global news and information concerning
En.wikipedia.org/wiki/Chicken	Biology, history, and information concerning the chicken and its care
Poultryhub.org	Australian resource centre for poultry events, announcements, news releases, job posting etc
Poultrypages.com	Provides the characteristics, history and details of many breeds of chickens, ducks, geese etc
Thegardencoop.com	Chicken coop plans and advice for do-it-yourself backyard poultry keepers
Msucares.com/poultry/consumer/	Research information on many topic areas in poultry husbandry by Mississippi State University
Poultryscience.org	Organization with the objective of advancing the poultry industry. Includes publications
Zootecnicainternational.com	Provides news concerning all the different sectors of the poultry industry
Guineafowl.com/fritsfarm/	Information on guinea fowl, raising guinea keets, incubation, housing, etc
Feathersite.com/Poultry/BRKPoultryPage.html	An on-line zoological garden of domestic poultry, including photos and information
Tulassi.com	Providers of poultry and agricultural management software for both egg and meat producing flocks
Quailsaustralia.com.au	Online community for people with an interest in quails, breeders and farmers
Dpi.nsw.gov.au/agriculture/livestock/poultry/species/duck-raising	Information about ducks, breeds and breeding, disease problems, housing, egg production, brooding etc
Chickscope.beckman.uiuc.edu	Students raise chicken embryos in the classroom and obtain magnetic

	resonance images
The-coop.org/forums/ubbthreads.php	Forum providing information on a variety of topics regarding the management and care of poultry
Dpi.nsw.gov.au/agriculture/livestock/poultry/species/geese-raising	Information about geese, breeds and breeding, housing, disease problems, handling and using geese etc
Poultryone.com	Resources for poultry husbandry and raising chickens such as links, poultry article archive etc
Ashtonwaterfowl.net/keeping_geese.htm	Provides answers to a number of common questions about the care of geese
Showsilkies.com	Articles on raising silkie chickens
Howtoraisequail.com	Information, articles and forum on raising quail

Source: alexa.com

Animal Breeding and Biotechnology

Website URL	Description
Iets.org	A group which serves as a professional forum for the exchange of information among scientists
Naab-css.org	Members are insemination services, semen sources, and research organizations
Bku.com/who.html	Describes services in breeding, and biotechnology in laboratory animals
Wiley.com/bw/journal.asp	Develops, publishes, and sells products in print and electronic media for educational, professional purposes
Cabi.org/agbiotechnet/	This site contains news, reviews, abstracts, reports, jobs, conferences and links on agriculture industry
Aaabg.org	A professional organisation based in Australia and New Zealand for livestock scientists, breeders etc
Fawc.org.uk/reports/dairy cow/dcowr048.htm	Recommendations on the procedures to be used when transferring embryos to recipient cattle
Cruachan.com.au/embryo _transfer.htm	Article by Ross Wilson explaining the concepts and practicalities of this procedure
Mcintoshembryo.com	Ontario's largest embryo transfer business
Sil.co.nz	Provides state of the art genetic information to New Zealand ram breeders

Cattle

Website URL	Description
Gala.com	A transgenics company using recombinant proteins for the improvement of livestock genetics
Beefstockerusa.org	Designed specifically for beef producers who background and/or run stocker or yearling cattle
Ianrpubs.unl.edu/epublic/pages/publicationD.jsp	University of Nebraska-Lincoln publications on feeding, breeding, herd management etc
Animalinfo.org/species/artiperi/bos_mutu.htm	Biology, ecology, habitat, and status of the yak, and information on its wild habitat
Cowbcs.info	Body condition scoring serves as a management tool to determine the nutritional needs of a cow

Source: alexa.com

Dairy

Website URL	Description
Nmconline.org	Not for profit organization that promotes research and provides information to the dairy industry
Ianrpubs.unl.edu/epublic/pages/index.jsp	University of Nebraska-Lincoln publications on feeding, breeding, herd management
Hoards.com	An online information resource for dairy farmers and anyone interested in the dairy industry
Adsa.org	An international organization of educators, scientists, and industrialists
Jds.fass.org	Official publication of the American Dairy Science Association
Dasc.vt.edu	Includes information on the department, and on goat milk production
Babcock.cals.wisc.edu	Information in more than 100 online technical documents about the dairy

	industry
Ars.usda.gov/main/site_main.htm	The main in-house research arm of the U.S. Department of Agriculture
Ianrpubs.unl.edu/epublic/pages/publicationD.jsp	University of Nebraska-Lincoln publications on feeding, breeding, herd management
Pods.dasnr.okstate.edu/docushare/dsweb/View/Collection-303	A list of publications on dairy production, as PDF files. From Oklahoma State University
Uwex.edu/milkquality	Information and resources related to the production of high quality milk
Brighton73.freeserve.co.uk/tomsplace/scientific/phd/4_survey/4_survey.htm	The results of a survey of dairy farmers on the use of High Dry-Matter silage
Britishmastitisconference.org.uk	An annual UK conference for anyone with an interest in dairy farming
Fawc.org.uk/reports/dairycow/dcowrtoc.htm	The Farm Animal Welfare Council provides recommendations on all aspects of the care of dairy
Milkacademy.com	This organisation aims to be a resource for sharing knowledge

Source: alexa.com

146

Pigs

Website URL	Description
Prairieswine.com	Conducts research on behalf of western Canadian pork producers
Dpi.nsw.gov.au/__data/assets/pdf_file/0019/56134/Basic_pig_husbandry-Gilts_and_sows_-_Primefact_70-final.pdf	Provides guidance on selection and care of gilts and sows, the mating, gestation etc
Dpi.nsw.gov.au/__data/assets/pdf_file/0003/56127/Basic_pig_husbandry_-_the_boar_-_Primefact_69-final.pdf	Provides guidance on selecting boars, mating management, feed requirements, disease control program etc
Dpi.nsw.gov.au/__data/assets/pdf_file/0019/56152/Basic_pig_husbandry-Grower_herd_-_Primefact_73-final.pdf	Provides guidance on feeding growing pigs, diseases, grading and temperature control etc
Dpi.nsw.gov.au/__data/assets/pdf_file/0018/56142/Basic_pig_husbandry-The_litter_-_Primefact_71-final.pdf	Provides information on managing newborn piglets, fostering and artificial rearing etc
Dpi.nsw.gov.au/__data/assets/pdf_file/0005/56147/Basic_pig_husbandry-The_weaner_-_Primefact_72-final.pdf	Discusses different types of weaning and their associated problems
Npip.une.edu.au	Provides a service to the Australian pig industry by providing genetic evaluation services
Thepigsite.com/pigjournal/	Produced by the UK Pig Veterinary Society
Epe.lac-bac.gc.ca/100/205/301/ic/cdc/hog/	Provides a historical account of the domestication of pigs and video clips of modern commercial production
Saskpork.com/pdfs/ed_pig_basics.pdf	Everything you ever needed to know about pigs but didn't know who to ask

Sheep

Website URL	Description
Grandin.com	Livestock behaviour, design of facilities and humane slaughter
Ramsem.com	A service for sheep breeders, linking the buyers and sellers of ram semen for AI purposes
Mylamb.org	A help community for people involved in 4H and FFA breeding sheep and market lamb projects
Nsip.org	NSIP offers a computerized genetic performance evaluation program
Flockfiler.com	FlockFiler is a computer database for keeping health, management, and breeding records of sheep
Agry.purdue.edu/ext/forages/	This site contains information on forages for the Midwestern U.S.
Ext.colostate.edu/pubs/livestk/01618.html	The Pearson square or box method of balancing rations is a simple procedure
Sheep.cornell.edu	Offers current information on management, nutrition, health, selection, and marketing for sheep
Ansci.cornell.edu/sheep/management/	Aims to evaluate economical methods of managing a highly productive flock to provide sheep
Grandin.com/references/new.corral.html	Information regarding livestock handling and stress reduction
Usbcha.com	News and events, club history, membership, and member information, dog profiles and photographs
Ext.colostate.edu/pubs/livestk/01608.html	Urea can be fed to ruminants as an economical replacement for a part of the protein in a ration
Ag.ansc.purdue.edu/sheep/articles/breeding.html	A webpage about ram rearing. A ram represents a sizeable investment
Ag.ansc.purdue.edu/sh	High hay prices are a major

eep/articles/highhay.ht ml	consideration for sheep producers.
Ag.ansc.purdue.edu/sh eep/articles/feeding.ht ml	What and how much should I feed my sheep?
Ag.ansc.purdue.edu/sh eep/articles/feedlamb. html	There are a number of ways to grow and finish lambs
Ag.ansc.purdue.edu/sh eep/articles/ideal.html	A webpage about ideal lambing season
Ag.ansc.purdue.edu/sh eep/articles/incrlamb.h tml	Proper management of the flock at key times of the production cycle can reduce young lamb mortality
Ag.ansc.purdue.edu/sh eep/articles/manewes. html	Separation at weaning can be stressful for both lambs and the ewes
Ag.ansc.purdue.edu/sh eep/articles/repromgt. html	Reproduction in sheep is influenced by numerous factors
Web2.canr.msu.edu/ha y/index.htm	This site, developed with the cooperation of the Michigan Hay and Grazing Council, Michigan
Dpi.nsw.gov.au/about us/resources/factsheet s/primefacts/	Official New South Wales government web site
Nal.usda.gov/awic/pu bs/livestock/lvstgran.h tm	Paper covering some of the factors that influence how an animal may react during handling
Nal.usda.gov/awic/pu bs/livestock/lvstshee.h tm	Bibliography and audiovisuals
Rurdev.usda.gov/rbs/c oops/cssheep.htm#ba cktable	An independent agency that assists the U.S. sheep and goat industries

TOP WEBSITES IN FISHERIES AND AQUACULTURE CATEGORY

Aquaculture

Website URL	Description
Thefishsite.com	TheFishSite.com provides free information
Gaalliance.org	An industry organization representing fish and shellfish farmers, particularly shrimp
Was.org	An international non-profit society founded in 1970
Aquaponics.net.au	Australian company offers information on growing vegetables and fish in a backyard aquaponics
Shrimpnews.com	Weekly news reports on issues related to world shrimp farming industry
En.wikipedia.org/wiki/Aquaculture	Information from Wikipedia on the cultivation of aquatic organisms under controlled conditions
Ag.arizona.edu/azaqua/	Information on aquaculture in Arizona, the United States
Nsgd.gso.uri.edu	Includes a digital library of resources on several aquaculture topics
Ag.auburn.edu/fish/	Academic and research programs for aquaculture and fisheries
Luciopercimprove.be	Research project at the University of Namur, Belgium, on improving pikeperch larval quality
Pdacrsp.oregonstate.edu	CRSP aims to advance science, research and education in aquatic resources
Aesweb.org	This organization aims to provide a forum for the discussion of engineering problems
Aquacultureassociation.ca	A non-profit charitable organization with the goals of fostering an aquaculture industry
Shellfish.org	An international organization of scientists, administrators and members of industry
Socalfishfarm.com	Aquaponics is a system that integrates aquaculture with hydroponics, the soil-free cultivation
En.wikipedia.org/w	Information from Wikipedia on this form

iki/Algaculture	of aquaculture involving the production of microalgae
Fao.org/fishery/culturedspecies/Litopenaeus_vannamei/en	Information from the FAO on the cultivation of this shrimp, its biology, habitat etc
En.wikipedia.org/wiki/Mariculture	Information from Wikipedia on this specialised form of aquaculture
Britishtrout.co.uk	Represents the UK trout farm industry and all aspects of trout research, promotion
Fao.org/fishery/culturedspecies/Macrobrachium_rosenbergii/en	Information from the FAO on this freshwater species, its biology, habitat, the production cycle etc
Fao.org/fishery/culturedspecies/Penaeus_monodon/en	Information from the FAO on this species, its biology, habitat, the production cycle etc
Cefas.co.uk/publications/marketing/crustaceans.pdf	CEFAS is a scientific research and consultancy centre providing services to aquaculturalists
Marindro.blogspot.com	Provides a discussion forum and information on shrimp and prawn culture in Indonesia
Fao.org/fishery/culturedspecies/Crassostrea_gigas/en	Information from the FAO on the culture of this species, its biology, habitat, the production cycle etc
Fao.org/fishery/culturedspecies/Oncorhynchus_mykiss/en	Information from the FAO on the cultivation of this species, its biology, habitat, etc

Fisheries

Website URL	Description
Tu.org	America's leading trout and salmon conservation organization
Dpi.nsw.gov.au/fisheries	Agency responsible for managing the fisheries resources of New South Wales
Fisheries.org	Professional society for fisheries scientists
Worldfishcenter.org	A food and environment research organization that joins forces with farmers, scientists, etc
Fish.washington.edu	Providing teaching and research in fisheries management and resource conservation
Nwfsc.noaa.gov	A research facility of the Northwest Region of the National Marine Fisheries Service, NOAA,
Psmfc.org	Promotes and supports policies and actions for conservation, development and management of fisheries
Billfish.org	Research, education, and advocacy to return billfish populations worldwide to healthy levels
Scotland.gov.uk/topics/marine	Marine laboratory, monitoring all aspects of marine, freshwater fisheries and environment
Fishwildlife.org	Quasi-official group of public agencies charged with protection and management of North America
Fao.org/fishery/en	World Food and Agriculture Organization's Major Programme on Fisheries
Pcouncil.org	Has developed fishery management plans for salmon, groundfish and coastal pelagic species
Fpc.org	Current and historic data on salmon and steelhead in the Snake and Columbia river basins
Asf.ca	Atlantic salmon conservation
Gcfi.org	GCFI is a not-for-profit organization that promotes the exchange of information
Seymoursalmon.c	Salmon enhancement and fisheries

om	education
Cbr.washington.edu	Efforts at the University of Washington to investigate issues surrounding salmon biology
Caribbeanfmc.com	U.S. government unit responsible for creating management plans for fishery resources (FMPs)
Fsbi.org.uk	Organization of professional fish biologists and fisheries managers in the British Isles
Iotc.org	Manages tuna and related species
Watershed-watch.org	Registered non-profit whose purpose is to conduct research, and educate the public
Ptagis.org	The Passive Integrated Transponder (PIT) tag has been developed as a research and management
Ifm.org.uk	The Institute of Fisheries Management (IFM) is an international organisation
Biotecmar.eu	Project to support the development of a chain for the production of valuable ingredients
Seafoodintelligence.com	Provides international market intelligence and a news service on freshwater and marine fisheries

Source: alexa.com

TOP WEBSITES IN FORESTRY CATEGORY

Forestry General

Website URL	Description
Arborday.org	Information on tree and shrub care; education resources related to trees
Arboristsite.com	A loose affiliation of volunteer tree working professionals that discuss the needs of tree care
Cifor.org	With a mission to contribute to the sustained well-being of people in developing countries
Esf.edu	Specialized college of the State University of New York located adjacent to Syracuse University
Environment.yale.edu	Professional and graduate school within Yale University
Fao.org/forestry/	Part of the Food and Agriculture Organization of the United Nations
Isa-arbor.com	Assisting tree care professionals in developing and maintaining effective plant health care program
Na.fs.fed.us	Site emphasis is on state and private forestry information
Forestryimages.org	Quality photographs of forest insects and disease organisms
Forestry.uga.edu	Include information about the forest business, their programs, alumni, and careers
Cfr.washington.edu	Offering programs in natural resource management and environmental sciences
Ffpri.affrc.go.jp	Focuses on sustainability, industry, rural community development, productivity
Nrs.fs.fed.us	Natural resource research and development in the Midwest
Cof.orst.edu	Contact and other information including links to the departments of Forest Engineering
Sfrc.ufl.edu	A department of the University of Florida
Kfri.res.in	Undertakes research and studies in

	forestry, bio-diversity, wildlife, wood and soil science
Bugwood.org	Clearing house to gather, create, maintain, promote the use of forestry
Srs.fs.usda.gov	Headquartered in Asheville, North Carolina, serves 13 Southern States and beyond
Fs.fed.us/psw/	Contains links to publications, journals, research proposals, databases and links to other USFS
Metla.fi	An independent research organization under the Ministry of Agriculture and Forestry
Foresthistory.org	Contains membership information, links to research and publications, quarterly journal etc
Ltrr.arizona.edu	Applies dendrochronology to understanding environmental variability
Forestry.ubc.ca	Undergraduate and graduate degree programs, as well as an international study program
Inbar.int	Knowledge dissemination for rural areas and industries on bamboo and rattan conservation and use
Efi.int	An independent non-governmental organisation conducting European forest research

Source: alexa.com

Arboriculture

Website URL	Description
Arboristsite.com	A loose affiliation of volunteer tree working professionals that discuss the needs of tree care
Isa-arbor.com	Assisting tree care professionals in developing and maintaining effective plant health care program
Trees.org.uk	Promotes care and knowledge of trees in the UK
Eac-arboriculture.com	A forum for arboricultural organizations in Europe
Home.ccil.org/~treeman/shigo/	A set of articles by Dr. Shigo on arboriculture, and on the growth of garden trees
Ohioline.osu.edu/hyg-fact/1000/1032.html	Provides guidance on selecting a consulting arborist, who offers advice
Aces.edu/pubs/docs/A/ANR-1255/ANR-1255.pdf	Copiously illustrated account of how to examine trees for safety, what symptoms to look for etc
Asiatreepreservation.com	ATP provides professional tree care, arboriculture training, and vegetation management consultancy
Cal-arb-association.com	Membership information, details and timetables of activities and workshops
Mac-isa.org	Information and links on trees and their care (arboriculture)

Bamboo

Website URL	Description
Inbar.int	Knowledge dissemination for rural areas and industries on bamboo and rattan conservation and use
Bamboo.org	Offers information about bamboo
Bamboocomposites.com	Describes the uses, species, conservation value and cultivation of bamboo in India
Bamboocentral.org	Promotes bamboo as an environmentally renewable resource
Illumin.usc.edu/article.php	Official site of the University of Southern California

Source: alexa.com

Silviculture

Website URL	Description
Acf.org/r_r.php	Project to introduce into the American chestnut the genetic material responsible for blight resistance
Rngr.net	RNGR aims to supply people who grow tree seedlings with the latest technical information
Ces.ncsu.edu/fletcher/programs/xmas/ctnotes/	A comprehensive series of informational sheets provided by North Carolina State University
Rngr.net/Publications/ttsm	Downloadable guide to seed biology, the collection, storage and germination of tropical tree
State.sc.us/forest/nur.htm	South Carolina Forestry Commission aims to provide high quality, improved seedlings
Ohioline.osu.edu/for-fact/0050.html	Crop trees are trees that produce the desired landowner benefits
Ohioline.osu.edu/for-fact/0048.html	Gives assistance in calculating the best time to harvest timber
Ohioline.osu.edu/for-fact/0034.html	Provides guidance on giving woodland proper care so that it remains healthy and vigorous
Edis.ifas.ufl.edu/pdffiles/EP/EP30000.pdf	Information on this process, typically requiring one or more prunings to develop a strong tree
Ces.ncsu.edu/nreos/forest/pdf/WON/won07.pdf	Provides guidance on calculating the site index
Ces.ncsu.edu/nreos/forest/pdf/WON/won02.pdf	Provides assistance in obtaining advice for forest landowners
Ces.ncsu.edu/nreos/forest/pdf/ag/ag-616.pdf	This forest landowners' guide gives assistance on assessing the risk of fire damage
Ces.ncsu.edu/nre	Outlines modifications to a planting plan

os/forest/pdf/W ON/won37.pdf	that can benefit wildlife, improve soil and water
Ces.ncsu.edu/nre os/forest/pdf/W ON/won34.pdf	Provides guidance on the removal of live or dead branches from standing trees
Ces.ncsu.edu/nre os/forest/pdf/W ON/won15.pdf	This article discusses the factors to be considered when deciding how to prepare the site
Ces.ncsu.edu/nre os/forest/pdf/W ON/won16.pdf	Provides guidance on drawing up a tree-planting contract, and help in selecting, handling, storage
Ces.ncsu.edu/nre os/forest/pdf/W ON/won13.pdf	Provides guidance on the cutting or removal of certain trees from a stand
Pubs.ext.vt.edu/4 65/465-315/465-315.pdf	Information on forest management and the shelterwood, seed-tree and clear-cutting systems
Warnell.forestry.u ga.edu/warnell/se rvice/library/inde x.php3	Include information about the forest business, their programs, alumni, and careers
Extension.missour i.edu/publications /DisplayPub.aspx:	Extends research and problem-solving resources from the University of Missouri

Urban Forestry

Website URL	Description
Arborday.org	Information on tree and shrub care; education resources related to trees
Menofthetrees.com.au	Non-profit organisation promoting community involvement in seed collecting, propagation, etc
Nrs.fs.fed.us/urban/	Information on current research on the effects of the urban forest on air quality, climate etc
Wane3000.com	Information and video about tree preservation methods in urban environments that provide water
Fs.fed.us/psw/programs/cufr/	Includes latest in urban forest research: GIS, inventory and monitoring, etc

Source: alexa.com

www.ingramcontent.com/pod-product-compliance
Lightning Source LLC
Chambersburg PA
CBHW070857180526
45168CB00005B/1849